AIM for Success Pr

Intermediate Algebra
An Applied Approach

NINTH EDITION

Richard N. Aufmann
Palomar College

Joanne S. Lockwood
Nashua Community College

Prepared by

Christine S. Verity

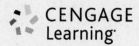

CENGAGE
Learning·

Australia • Brazil • Mexico • Singapore • United Kingdom • United States

For product information and technology assistance, contact us at
Cengage Learning Customer & Sales Support, 1-800-354-9706.

For permission to use material from this text or product, submit all requests online at **www.cengage.com/permissions**
Further permissions questions can be emailed to **permissionrequest@cengage.com**.

ISBN-13: 978-1-285-42024-0
ISBN-10: 1-285-42024-1

Cengage Learning
200 First Stamford Place, 4th Floor
Stamford, CT 06902
USA

Cengage Learning is a leading provider of customized learning solutions with office locations around the globe, including Singapore, the United Kingdom, Australia, Mexico, Brazil, and Japan. Locate your local office at: **www.cengage.com/global**.

Cengage Learning products are represented in Canada by Nelson Education, Ltd.

To learn more about Cengage Learning Solutions, visit **www.cengage.com**.

Purchase any of our products at your local college store or at our preferred online store **www.cengagebrain.com**.

Printed in the United States of America
2 3 4 5 6 7 19 18 17 16 15

Contents

Name   Score

Find the Additive Inverse.

1. −3   2. −7   1. _____

                                                                2. _____

3. 5   4. 17   3. _____

                                                                4. _____

5. −87   6. 72   5. _____

                                                                6. _____

7. 66   8. 52   7. _____

                                                                8. _____

9. −120   10. −36   9. _____

                                                                10. _____

Solve.

11. Let $x \in \{-9, -3, 4\}$. For which values   12. Let $y \in \{-8, -2, 1, 6\}$. For which values   11. _____
    of x is $x \geq -3$ true?   of x is $y < 1$ true?
                                                                                                   12. _____

13. Let $a \in \{-60, 13, 28\}$. Evaluate $-a$ 14. Let $a \in \{-48, -5, 9\}$. Evaluate $|b|$ 13. _____
    for each element of the set.   for each element of the set.
                                                                                                   14. _____

1

Name _____   Score _____

Use the roster method to write the set.

1. the integers between −3 and 2   2. the odd natural numbers less than 9   1. _____

                                                                                               2. _____

3. the positive integer multiples of 5   4. the negative integer multiples of 6   3. _____
    that are less than or equal to 35   that are greater than −23
                                                                                               4. _____

Use set builder notation to write the set.

5. the real numbers greater than or   6. the real numbers between −11 and −2   5. _____
    equal to −6
                                                                                               6. _____

7. the real numbers less than 10   8. the real numbers between −1 and   7. _____
                                               1, inclusive
                                                                                               8. _____

Graph.

9. $\{x | x \le 2\}$ 10. $\{x | x \ge -3\}$ 9. _____

    -5 -4 -3 -2 -1 0 1 2 3 4 5   -5 -4 -3 -2 -1 0 1 2 3 4 5   10. _____

11. $\{x | x < 0\}$ 12. $\{x | -4 \le x < 4\}$ 11. _____

    -5 -4 -3 -2 -1 0 1 2 3 4 5   -5 -4 -3 -2 -1 0 1 2 3 4 5   12. _____

Write each set of real numbers in interval notation.

13. $\{x | x \ge -3\}$ 14. $\{x | -4 < x \le -2\}$ 13. _____

                                                                                              14. _____

Write each set of real numbers in set-builder notation.

15. $(-\infty, -1)$ 16. $[4, 7)$ 15. _____

                                                                                              16. _____

Graph.

17. $[0, 1)$ 18. $(-\infty, 0)$ 17. _____

                                                                                              18. _____

19. $(2, 4]$ 20. $[-1, 2)$ 19. _____

                                                                                              20. _____

2

Name _____   Score _____

Find $A \cup B$.

1. $A = \{-6, 0, 6\}$,
    $B = \{-1, 0, 1\}$

2. $A = \{-3, -1, 0, 1\}$,
    $B = \{0, 1, 3\}$

3. $A = \{-2, -1, 0\}$,
    $B = \{-1, 0, 2\}$

1. _____

2. _____

3. _____

Find $A \cap B$.

4. $A = \{-6, 0, 6\}$,
    $B = \{-2, 0, 2\}$

5. $A = \{1, 2, 5, 10\}$,
    $B = \{5, 10, 50\}$

6. $A = \{1, 4, 9, 16\}$,
    $B = \{1, 4, 9\}$

4. _____

5. _____

6. _____

Graph the solution set.

7. $\{x | x > -2\} \cap \{x | x < 2\}$
    -5 -4 -3 -2 -1 0 1 2 3 4 5

8. $\{x | x \geq -4\} \cap \{x | x \leq 3\}$
    -5 -4 -3 -2 -1 0 1 2 3 4 5

7. _____

8. _____

9. $\{x | x < 1\} \cap \{x | x > -1\}$
    -5 -4 -3 -2 -1 0 1 2 3 4 5

10. $\{x | x < -1\} \cup \{x | x > 1\}$
    -5 -4 -3 -2 -1 0 1 2 3 4 5

9. _____

10. _____

11. $\{x | x \leq -1\} \cup \{x | x > 0\}$
    -5 -4 -3 -2 -1 0 1 2 3 4 5

12. $\{x | x > 0\} \cup \{x | x > 2\}$
    -5 -4 -3 -2 -1 0 1 2 3 4 5

11. _____

12. _____

13. $[0, 2] \cap [1, 4]$
    -5 -4 -3 -2 -1 0 1 2 3 4 5

14. $(2, 3) \cup (4, \infty)$
    -5 -4 -3 -2 -1 0 1 2 3 4 5

13. _____

14. _____

15. $[3, \infty) \cup (-\infty, -3]$
    -5 -4 -3 -2 -1 0 1 2 3 4 5

16. $(-\infty, -2) \cap (-\infty, 3]$
    -5 -4 -3 -2 -1 0 1 2 3 4 5

15. _____

16. _____

17. $(3, \infty) \cup (-3, \infty)$
    -5 -4 -3 -2 -1 0 1 2 3 4 5

18. $(0, \infty) \cup (-5, \infty)$
    -5 -4 -3 -2 -1 0 1 2 3 4 5

17. _____

18. _____

Name   Score

Simplify.

1. $-14 + (-12)$ 2. $-18 - 9$ 3. $8 - 17$

4. $8 \cdot (-30)$ 5. $2 \cdot 5 \cdot (-6)$ 6. $15 \cdot 0 \cdot (-3)$

7. $18 \div (-3)$ 8. $-30 \div (6)$ 9. $-54 \div (-8)$

10. $(-3)(2)(-4)(5)$ 11. $(-5)(6)(-13)$ 12. $(27)(12)(-50)$

13. $-176(-215)$ 14. $|(-10)(8)|$ 15. $|14(-6)|$

16. $|6 - 10|$ 17. $|12 - (-8)|$ 18. $|-12 - (-4)|$

19. $|-16 + 12|$ 20. $|-46 \div 6|$ 21. $|-52 \div (-4)|$

22. $-3 + (-17) - 15 + 9$ 23. $-5 + (-8) - 14 + 26$ 24. $6 - (-14) + 10 - 8$

25. $5 - |7 - 11|$ 26. $-2 - |-2 - (-3)|$ 27. $-10 - |-6 - (-14)|$

1. _____

2. _____

3. _____

4. _____

5. _____

6. _____

7. _____

8. _____

9. _____

10. _____

11. _____

12. _____

13. _____

14. _____

15. _____

16. _____

17. _____

18. _____

19. _____

20. _____

21. _____

22. _____

23. _____

24. _____

25. _____

26. _____

27. _____

Name   Score

Evaluate.

1. 2^4 2. -5^3 3. $(-4)^3$ 1. _____

                                                                          2. _____

                                                                          3. _____

4. $(-3)^4$ 5. $(-7)^2$ 6. $(-3)^5$ 4. _____

                                                                          5. _____

                                                                          6. _____

7. -9^2 8. -5^3 9. $(-4)^4$ 7. _____

                                                                          8. _____

                                                                          9. _____

10. $2^2 \cdot 3^2$ 11. $(-2)^2(3)^2$ 12. $(-2)^4(-4)^2$ 10. _____

                                                                          11. _____

                                                                          12. _____

13. $2^2 \cdot 3^3$ 14. $-3^2 \cdot 2^3$ 15. $3^2 \cdot 4^3$ 13. _____

                                                                          14. _____

                                                                          15. _____

16. $-2^3 \cdot 3^2$ 17. $(-3)^2(-2)^3$ 18. $(-3)^2 \cdot (-5)^2$ 16. _____

                                                                          17. _____

                                                                          18. _____

19. $-5^3 \cdot 4^2$ 20. $(-2)^2(-3)^3$ 21. $(-3)^2(-2)^4$ 19. _____

                                                                          20. _____

                                                                          21. _____

22. $(-5^2)(-4)^2$ 23. $(-5)^3(-4)^3$ 24. $-2^2 \cdot (-3)^2 \cdot 4$ 22. _____

                                                                          23. _____

                                                                          24. _____

25. $-5(-2)^2(2^3)$ 26. $-5(2^2)(-2)^3$ 27. $2 \cdot 3^2 \cdot 4^2$ 25. _____

                                                                          26. _____

                                                                          27. _____

Name   Score

Simplify.

1. $\dfrac{3}{8} + \dfrac{5}{6}$

2. $\dfrac{5}{16} - \dfrac{7}{12}$

3. $-\dfrac{4}{5} - \dfrac{11}{15}$

4. $\dfrac{3}{5} + \dfrac{1}{10} - \dfrac{9}{20}$

5. $\dfrac{2}{3} + \dfrac{17}{24} - \dfrac{5}{6}$

6. $\dfrac{3}{5} - \dfrac{9}{10} + \dfrac{11}{30}$

7. $\dfrac{1}{8} - \dfrac{17}{32} + \dfrac{3}{4}$

8. $-\dfrac{1}{2} \cdot \dfrac{5}{6}$

9. $\dfrac{3}{4}\left(-\dfrac{8}{15}\right)\dfrac{25}{32}$

10. $-\dfrac{9}{20} \div \dfrac{3}{5}$

11. $-\dfrac{4}{9} \div \left(-\dfrac{16}{36}\right)$

12. $\dfrac{8}{15} \div \left(-\dfrac{32}{45}\right)$

13. $\left(-\dfrac{3}{4}\right)\left(\dfrac{5}{18}\right)\left(\dfrac{6}{25}\right)$

14. $\dfrac{25}{27}\left(-\dfrac{7}{20}\right)\left(-\dfrac{18}{35}\right)$

15. $\dfrac{9}{16} \div \left(-\dfrac{27}{32}\right)$

16. $-16.72 + (-8.91)$

17. $-25.6 + 16.04$

18. $-10.36 + 2.384$

19. $-0.161 + 138.2$

20. $2.346 - 7.85$

21. $(2.86)(3.5)$

22. $(0.02)(8.6)(5.5)$

23. $(0.6)(2.3)(-3.5)$

24. $-0.5655 \div (-0.13)$

25. $-0.325 \div 0.065$

26. $-5.47 - 2.86 + 0.764$

27. $7.7 + 2.164 - 36.346$

1. _____
2. _____
3. _____
4. _____
5. _____
6. _____
7. _____
8. _____
9. _____
10. _____
11. _____
12. _____
13. _____
14. _____
15. _____
16. _____
17. _____
18. _____
19. _____
20. _____
21. _____
22. _____
23. _____
24. _____
25. _____
26. _____
27. _____

Name _____ Score _____

Simplify.

1. $8-3(8\div 4)^2$

2. $5^2-(8-3)^2\cdot 3$

1. _____

2. _____

3. $18-\dfrac{3^2-5}{2^3+2}$

4. $\dfrac{3(6-4)}{6^2-5^2}$

3. _____

4. _____

5. $\dfrac{3+\dfrac{5}{6}}{\dfrac{7}{18}}$

6. $\dfrac{\dfrac{5}{27}}{4-\dfrac{8}{9}}$

5. _____

6. _____

7. $4[(3-7)\cdot 5-3]$

8. $3[(12\div 3)-(-3)]+5$

7. _____

8. _____

9. $18-3\left(\dfrac{12-3}{2-5}\right)\div\dfrac{1}{3}$

10. $30\div 6\left(\dfrac{15+5}{-2^2+9}\right)-3$

9. _____

10. _____

11. $6[4-(-5+8)\div 3]$

12. $16-4[5-(-2+8)-9]$

11. _____

12. _____

13. $\dfrac{2}{3}-\left(\dfrac{3}{4}\div\dfrac{9}{16}\right)+\dfrac{7}{12}$

14. $\left(-\dfrac{3}{4}\right)^2-\dfrac{7}{8}\cdot\dfrac{8}{21}+\dfrac{11}{12}$

13. _____

14. _____

15. $\dfrac{2}{3}-\dfrac{\dfrac{13}{24}}{3-\dfrac{1}{3}}\div\dfrac{1}{8}$

16. $\dfrac{5}{6}+\dfrac{2-\dfrac{2}{3}}{\dfrac{8}{9}}\cdot\dfrac{2}{3}$

15. _____

16. _____

17. $0.3(1.3-2.1)^2+4.7$

18. $4.2-(0.4)^2\div 0.08$

17. _____

18. _____

Name   Score

Use the given Property of the Real Numbers to complete the statement.

1. The Associative Property of Addition   2. The Commutative Property of   1. _____
    $(7+3)+5 = ?+(3+5)$ Multiplication
                                                      $8 \cdot 5 = 5 \cdot ?$ 2. _____

3. The Multiplication Property of One   4. The Distributive Property   3. _____
    $7 \cdot ? = 7$ $(8+3)7 = 8 \cdot 7 + ?$
                                                                                           4. _____

5. The Inverse Property of Addition   6. The Division Properties of Zero   5. _____
    $x + ? = 0$ $\dfrac{?}{5} = 0$
                                                                                           6. _____

7. The Multiplication Property of Zero   8. The Commutative Property of Addition 7. _____
    $6 \cdot 0 = ?$ $xy + wz = ?+ xy$
                                                                                           8. _____

9. The Associative Property of Multiplication   10. The Inverse Property of Multiplication 9. _____
    $4(5x) = ? \cdot x$ $\dfrac{1}{x}(x) = ?$
                                                                                           10. _____

11. The Addition Property of Zero   12. The Distributive Property   11. _____
    $4 + ? = 4$ $3(x+2) = ? \cdot x + 6$
                                                                                           12. _____

Identify the property that justifies the statement.

13. $-11+11 = 0$ 14. $d(f+g) = df + dg$ 13. _____

                                                                                           14. _____

15. $(-45y)(0) = 0$ 16. $a \cdot 1 = a$ 15. _____

                                                                                           16. _____

17. $(a+b)+c = a+(b+c)$ 18. $\dfrac{0}{7} = 0$ 17. _____

                                                                                           18. _____

19. $\dfrac{7}{0}$ is undefined   20. $-7+0 = -7$ 19. _____

                                                                                           20. _____

21. $a(bc) = (ab)c$ 22. $mn\left(\dfrac{1}{mn}\right) = 1$ 21. _____

                                                                                           22. _____

23. $a+b = b+a$ 24. $(a \cdot b) \cdot c = c \cdot (a \cdot b)$ 23. _____

                                                                                           24. _____

Name   Score

Evaluate the variable expression when $a = 1$, $b = -2$, $c = 3$, and $d = -4$.

1. $ad + bc$

2. $d^2 + (a-b)^2$

3. $(c-3a)^2 + b$

4. $(2a-d)^2 \div 3b$

5. $c^2 - (a-d)^2$

6. $(c-3b)^2 + d$

7. $2(a+b)^2 \div d^2$

8. $3b^2 \div \dfrac{b-d}{2}$

9. $\dfrac{2bc}{-3} + a^2$

10. $\dfrac{3a+3c}{d}$

11. $\dfrac{2d-b}{b-2a}$

12. $\dfrac{a-c}{d-b}$

13. $\left| b^2 + 2d \right|$

14. $-c \left| a + 2b \right|$

15. $b \left| c - 2d \right|$

16. $b^2 - a^2$

17. $a^2 + b^2 - c^2$

18. $\left| d^2 - b^2 \right|$

19. $-b \left| c + 2d \right|$

20. $\dfrac{2d-a}{c-2a}$

21. $\dfrac{3c+a^2}{b^2+a}$

22. $\dfrac{d^2 - 2c}{a^3 + b^2}$

23. $\dfrac{c^2 + a^2}{d - b}$

24. $\dfrac{a^2 - d^2}{a + d}$

25. $-3bc - \left| \dfrac{ac-b}{ab+c} \right|$

26. $\dfrac{3(d+b)}{c-2a}$

27. $(d+2b) \div a^3$

1. _____

2. _____

3. _____

4. _____

5. _____

6. _____

7. _____

8. _____

9. _____

10. _____

11. _____

12. _____

13. _____

14. _____

15. _____

16. _____

17. _____

18. _____

19. _____

20. _____

21. _____

22. _____

23. _____

24. _____

25. _____

26. _____

27. _____

Name   Score

Simplify.

1. $6x + 5x$ 2. $2x + 12x$ 3. $-3x + 7x - 9x$ 1. _____

                                                                        2. _____

                                                                        3. _____

4. $4x - 7x + 10x$ 5. $-4b + 8a + 9b$ 6. $2a - 9b - 11a$ 4. _____

                                                                        5. _____

                                                                        6. _____

7. $10\left(\dfrac{1}{10}y\right)$ 8. $\dfrac{1}{5}(5x)$ 9. $-(x - y)$ 7. _____

                                                                        8. _____

                                                                        9. _____

10. $-(-x + y)$ 11. $2(-3a + a - 6)$ 12. $4x - 3(5x - 6)$ 10. _____

                                                                        11. _____

                                                                        12. _____

13. $5x - 3(x - 8y)$ 14. $5[a - 4(4 - 5a)]$ 15. $4[x - 3(x - 2y)]$ 13. _____

                                                                        14. _____

                                                                        15. _____

16. $-6(x + 2y) - 3x$ 17. $2[x - 4(x + 3y)]$ 18. $4[y - 2(y + 3x)]$ 16. _____

                                                                        17. _____

                                                                        18. _____

19. $-3(x - 2y) + 5(3x - 4y)$ 20. $4(2a - 3b) - 5(-4a + 5b)$ 21. $-6(3a - b) + 5(-2b + a)$ 19. _____

                                                                        20. _____

                                                                        21. _____

22. $6x - 5[y - 3(x - 2[x - 2y])]$ 23. $3x - 2[x - 3(y - [4y + 5])]$ 24. $5 - 3(5x - 3y) - 2(-5x + 4y)$ 22. _____

                                                                        23. _____

                                                                        24. _____

25. $2x + 7(x - 2) - 2(2x - 3)$ 26. $\dfrac{1}{5}[19x - 4(x - 10) + 5]$ 27. $\dfrac{1}{2}[15x - 3(x - 4) - 6x]$ 25. _____

                                                                        26. _____

                                                                        27. _____

Name

Score

Translate into a variable expression. Then simplify.

1. a number plus twice a number

2. a number increased by the difference between six and the number

3. the difference between a number and one third of the number

4. one third of the total of nine times a number and twenty-one

5. nine times the quotient of a number and three

6. the sum of one half of a number and three fourths of the number

7. twelve times the product of two and a number

8. a number decreased by three fourths of the number

9. a number minus the difference of the number and four

10. the total of four times a number and twice the difference between the number and four

11. twice the total of two consecutive odd integers

12. twenty minus one third of a number and nine

13. the sum of the second and third of three consecutive even integers

14. one third of the sum of three consecutive odd integers

15. one less than the sum of the first and third of three consecutive integers

16. seven more than twice the sum of a number and eleven

17. the sum of four times a number and six subtracted from nine times the number

18. four plus the product of five more than a number and nine

19. a number subtracted from the product of six plus the number and ten

20. eleven decreased by the product of two less than a number and ten

1. _____

2. _____

3. _____

4. _____

5. _____

6. _____

7. _____

8. _____

9. _____

10. _____

11. _____

12. _____

13. _____

14. _____

15. _____

16. _____

17. _____

18. _____

19. _____

20. _____

Name   Score

Solve.

1. A 10-foot board is cut into two different lengths. Express the length of the longer piece in terms of the length of the shorter piece.

2. The measure of angle B of a triangle is five times the measure of angle A. The measure of angle C is ten times the measure of angle B. Write the expressions for angle B and angle C in terms of angle A.

1. _____

2. _____

3. A fishing line 3 m long is cut into two pieces, one longer than the other. Express the length of the longer piece in terms of the length of the shorter piece.

4. A financial advisor has invested $20,000 in two accounts. If one account contains d dollars, express the amount in the second account in terms of d.

3. _____

4. _____

5. A trail mixture contains 2 lb more of raisins than peanuts. Express the amount of raisins in terms of the amount of peanuts in the mixture.

6. The Ambassador Bridge from Detroit to Canada is 1883 m longer than the Detroit – New Windsor Tunnel. Express the length of the bridge in terms of the length of the tunnel.

5. _____

6. _____

7. A wallet contains four times as many $20 dollar bills as single dollar bills. Express the number of $20 dollar bills in terms of the number of singles.

8. The number of gallons of white paint is twice the number of gallons of yellow paint. Express the number of gallons of white paint in terms of the number of gallons of yellow paint.

7. _____

8. _____

Name	Score

Solve and check.

1. $x - 3 = 6$ **2.** $x - 7 = 3$ **3.** $b - 2 = -7$

4. $c - 7 = -1$ **5.** $8 + m = 4$ **6.** $7 + m = -2$

7. $4x = 20$ **8.** $6x = 3$ **9.** $-2x = 5$

10. $-3a = 7$ **11.** $-9 + x = 12$ **12.** $\dfrac{3}{8} + x = \dfrac{15}{16}$

13. $x + \dfrac{3}{5} = \dfrac{11}{15}$ **14.** $\dfrac{9}{10} - y = \dfrac{4}{5}$ **15.** $\dfrac{2}{3}y = 8$

16. $\dfrac{4}{5}y = 8$ **17.** $\dfrac{5t}{6} = -10$ **18.** $\dfrac{4a}{9} = -36$

19. $-\dfrac{5}{6}x = \dfrac{3}{5}$ **20.** $-\dfrac{7}{8}y = \dfrac{5}{12}$ **21.** $-\dfrac{9x}{10} = \dfrac{1}{4}$

22. $-\dfrac{3}{4}x = -\dfrac{9}{16}$ **23.** $\dfrac{2a}{7} = -14$ **24.** $\dfrac{3t}{10} = -6$

25. $x + 9.75 = 13$ **26.** $x - 16.74 = -1.36$ **27.** $x + 2.76 = -1.08$

1.	_____
2.	_____
3.	_____
4.	_____
5.	_____
6.	_____
7.	_____
8.	_____
9.	_____
10.	_____
11.	_____
12.	_____
13.	_____
14.	_____
15.	_____
16.	_____
17.	_____
18.	_____
19.	_____
20.	_____
21.	_____
22.	_____
23.	_____
24.	_____
25.	_____
26.	_____
27.	_____

Name   Score

Solve and check.

1. $2x + 3x = 15$ **2.** $3x - 7x = 16$ **3.** $6x + 8 = 2$

1. _____

2. _____

3. _____

4. $2x - 3x = 0$ **5.** $3a - 4a = 9 - 4a$ **6.** $8 - 3t = 2t + 18$

4. _____

5. _____

6. _____

7. $6x - 10 = 3x + 5$ **8.** $1 - 4t = t + 16$ **9.** $7 - 6t = 3t + 25$

7. _____

8. _____

9. _____

10. $2a - 7a = 9 - 6a$ **11.** $4a + 7a = 8a - 9$ **12.** $2b - b = 5 - 2b$

10. _____

11. _____

12. _____

13. $5x - 2 = 6 + x$ **14.** $2x + 8 = 3 + 7x$ **15.** $\frac{3}{4}y + y = 14$

13. _____

14. _____

15. _____

16. $\frac{5}{6}y - 3 = 7$ **17.** $\frac{3}{10}x + 2 = 3$ **18.** $4x - 3x + 6 = 11 - 4x$

16. _____

17. _____

18. _____

19. $x - 8x + 2 = 5 - 4x$ **20.** $6 + 5y - 11 = 2y - 7$ **21.** $-3t + 7 - t = -7t - 5$

19. _____

20. _____

21. _____

22. $2x - 7 + 6x = 17 + 4x + x$ **23.** $3x - 7x + 5 = 3 - 5x$ **24.** $4t - 6 + 6t = 18 + 2t$

22. _____

23. _____

24. _____

25. $7 + 5x - 13 = -2x + 6x + 2$ **26.** $3x - 5 + 7x = 21 - 3x$ **27.** $8 + 5x - 10 = -2x + 6x + 12$ 25. _____

26. _____

27. _____

Name _____   Score _____

Solve and check.

1. $3(2x-1)-4x=9+4x$ 2. $10-2x=8x-5(4-x)$ 1. _____

                                                                                2. _____

3. $5(x-1)+3=3x-2(3-2x)$ 4. $14-4x=6x-2(5-x)$ 3. _____

                                                                                4. _____

5. $3(x-2)+6=3x-2(3-x)$ 6. $5x-2(x-1)=3(4-2x)+8$ 5. _____

                                                                                6. _____

7. $6(x-1)+3=3x-2(1-x)$ 8. $3x-4(x+2)=3(2-x)+4$ 7. _____

                                                                                8. _____

9. $-4(x-1)=3[x-2(x-4)+x]$ 10. $2[x-(3-x)-3x]=4(5-x)$ 9. _____

                                                                                10. _____

11. $\dfrac{3}{8}t-\dfrac{3}{4}=\dfrac{1}{2}t$ 12. $\dfrac{2}{3}t-\dfrac{1}{6}t=2$ 11. _____

                                                                                12. _____

13. $\dfrac{3}{4}x-\dfrac{2}{3}x-4=\dfrac{1}{6}x-7$ 14. $\dfrac{2}{3}x-\dfrac{1}{6}x+\dfrac{1}{3}=\dfrac{2}{3}x-\dfrac{3}{2}$ 13. _____

                                                                                14. _____

15. $\dfrac{4x-5}{3}-3x=10$ 16. $\dfrac{2a-7}{5}+3=2a$ 15. _____

                                                                                16. _____

17. $\dfrac{2x-8}{2}-2x=10$ 18. $\dfrac{3a-7}{4}+2=a$ 17. _____

                                                                                18. _____

19. $\dfrac{x-1}{3}-\dfrac{x+4}{6}=\dfrac{4x-3}{2}$ 20. $\dfrac{3x+1}{3}+\dfrac{2x-3}{4}=\dfrac{3x-2}{6}$ 19. _____

                                                                                20. _____

15

Name   Score

Rewrite the formula in terms of the variable given.

1. $a_n = a_1 + (n-1)d;\ n$ 2. $A = \frac{1}{2}bh;\ b$

3. $x = -\frac{b}{2a};\ a$ 4. $A = P + \text{Pr}\,t;\ t$

5. $PV = nRT;\ n$ 6. $A = \frac{1}{2}bh;\ h$

7. $F = \frac{Gm_1 m_2}{r^2};\ G$ 8. $\frac{P_1 V_1}{T_1} = \frac{P_2 V_2}{T_2};\ V_2$

9. $I = \frac{E}{R+r};\ R$ 10. $A = \frac{1}{2}h(b_1 + b_2);\ b_1$

11. $d = rt;\ r$ 12. $I = \frac{100M}{C};\ C$

13. $E = aI(T - t);\ T$ 14. $\frac{1}{t} = \frac{1}{a} + \frac{1}{b};\ a$

15. $d = \frac{a}{2}(2t - v);\ v$ 16. $P = \frac{R - C}{n};\ n$

17. $S = 2wh + 2wL + 2Lh;\ w$ 18. $h = vt + \frac{1}{2}gt^2;\ g$

1. _____

2. _____

3. _____

4. _____

5. _____

6. _____

7. _____

8. _____

9. _____

10. _____

11. _____

12. _____

13. _____

14. _____

15. _____

16. _____

17. _____

18. _____

Name _____   Score _____

Solve.

1. A coffee merchant wants to make 45 lb of a blend of coffee to sell for $5.80 per pound. The blend is made by using a $7 grade and a $5 grade of coffee. How many pounds of each grade of coffee should be used?

2. A butcher combined 100 lb of hamburger that cost $1.75 per pound with 40 lb of hamburger that cost $3.50 per pound. Find the selling price of the hamburger mixture.

1. _____

2. _____

3. Tickets for a "little theater" production sold for $4.00 for each adult and $1.50 for each child. The total receipts for 400 tickets sold were $1400. Find the number of adult tickets sold.

4. For one performance of a play, 480 tickets were sold. Adult tickets sold for $3.50 each, and children's tickets sold for $1.00 each. Receipts from the sale of the tickets totaled $1230. Find the number of adult tickets sold.

3. _____

4. _____

5. To make a flour mix, a miller combines soybeans that cost $7.25 per bushel with wheat that cost $4.75 per bushel. How many bushels of each were used to make a mixture of 1000 bushels to sell for $5.25 per bushel?

6. Forty liters of pure maple syrup that cost $9.00 per liter were mixed with imitation maple syrup that cost $4.25 per liter. How much imitation maple syrup is needed to make a mixture to sell for $5.50 per liter?

5. _____

6. _____

7. How many ounces of pure silver that cost $11.50 an ounce must be mixed with 80 oz of a silver alloy that cost $7.75 an ounce to make an alloy that cost $8.50 an ounce?

8. A goldsmith combined pure gold that cost $420 per ounce with an alloy of gold that cost $180 per ounce. How many ounces of each were used to make 60 oz of gold alloy to sell for $270 per ounce?

7. _____

8. _____

9. Find the selling price per ounce of a face cream mixture made from 80 oz of face cream that cost $3.60 per ounce and 40 oz of face cream that cost $12.60 per ounce.

10. Find the selling price per pound of a tea mixture made from 50 lb of tea that cost $5.70 per pound and 90 lb of tea that cost $3.60 per pound.

9. _____

10. _____

11. How many pounds of peanuts that cost $2.00 per pound must be mixed with 50 lb of cashews that cost $5.50 per pound to make a mixture that costs $3.25 per pound?

12. Cranberry juice that cost $4.50 per gallon was mixed with 60 gal of apple juice that cost $2.25 per gallon. How much cranberry juice was used to make cran-apple juice to sell for $3.00 per gallon?

11. _____

12. _____

17

Name _____   Score _____

Solve.

1. A silversmith mixed 40 g of a 70% silver alloy with 60 g of a 30% silver alloy. What is the percent concentration of the resulting alloy?

2. A goldsmith mixed 20 g of a 60% gold alloy with 80 g of a 20% gold alloy. What is the percent concentration of the resulting alloy?

3. How many milliliters of pure acid must be added to 60 ml of a 30% acid solution to make a 50% acid solution?

4. How many ounces of pure water must be added to 75 oz of a 10% salt solution to make a 6% salt solution?

5. A butcher has some hamburger that is 24% fat and some hamburger that is 16% fat. How many pounds of each should be mixed to make 80 lb of hamburger that is 18% fat?

6. A hospital staff mixed a 70% disinfectant solution with a 30% disinfectant solution. How many liters of each were used to make 30L of a 45% disinfectant solution?

7. How many grams of a 4% salt solution must be mixed with 50 g of a 9% salt solution to make a 6% salt solution?

8. How many pounds of a 12% aluminum alloy must be mixed with 400 lb of a 25% aluminum alloy to make a 17% aluminum alloy?

9. A chemist mixed a 70% alcohol solution with a 30% alcohol solution to make a 45% alcohol solution. How many liters of each were used to make 120 L of a 45% solution?

10. An alloy containing 20% tin is mixed with an alloy containing 80% tin. How many pounds of each were used to make 600 lb of an alloy containing 35% tin?

11. How many milliliters of a 10% solution of boric acid should be mixed with 60 ml of a 4% boric acid solution to produce a 6% boric acid solution?

12. How many milliliters of pure alcohol should be mixed with 300 ml of a 25% alcohol solution to product a 40% alcohol solution?

1. _____

2. _____

3. _____

4. _____

5. _____

6. _____

7. _____

8. _____

9. _____

10. _____

11. _____

12. _____

Name _____   Score _____

Solve.

1. Two buses start from the same station and drive in opposite directions. The express bus is traveling 15 mph faster than the local bus. In 3 h the buses are 345 mi apart. Find the rate of each bus.

2. Two planes start from the same point and fly in opposite directions. The first plane is flying 30 mph slower than the second plane. In 2 h the planes are 1100 mi apart. Find the rate of each plane.

1. _____

2. _____

3. A speeding car going 70 mph has a 1-hour head start on a helicopter trying to overtake the car. The helicopter is traveling at 140 mph. How far from the starting point does the helicopter overtake the car?

4. A car traveling 52 mph overtakes a cyclist who, riding at 12 mph, has a 5-hour head start. How far from the starting point does the car overtake the cyclist?

3. _____

4. _____

5. A commuter plane flies to a small town from a major airport. The average speed flying to the small town was 280 mph, and the average speed returning was 200 mph. The total flying time was 3 h. Find the distance between the two airports.

6. A cabin cruiser left a harbor and traveled to a small island at an average speed of 27 mph. On the return trip, the cabin cruiser traveled at an average speed of 18 mph. The total time for the trip was 5 h. How far was the island from the harbor?

5. _____

6. _____

7. Two cars are 318 mi apart and traveling toward each other. One car travels 6 mph faster than the other car. The cars meet in 3 h. Find the speed of each car.

8. Two planes are 1500 mi apart and traveling toward each other. One plane is traveling 40 mph faster than the other plane. The planes meet in 2.5 h. Find the speed of each plane.

7. _____

8. _____

9. Two planes start from the same point and fly in opposite directions. The first plane is flying 40 mph faster than the second plane. In 3 h the planes are 960 mi apart. Find the speed of each plane.

10. An executive has an appointment 1340 mi from the office. The executive takes a helicopter from the office to the airport and a plane from the airport to the business appointment. The helicopter averages 80 mph and the plane averages 520 mph. The total time spent traveling is 3 h. Find the distance from the executive's office to the airport.

9. _____

10. _____

Name   Score

Solve.

1. $x - 4 < 2$ 2. $x + 5 \geq 2$ 1. _____

                                                                        2. _____

3. $2x \leq 10$ 4. $3x > 15$ 3. _____

                                                                        4. _____

5. $-4x > 12$ 6. $-5x \leq -20$ 5. _____

                                                                        6. _____

7. $7x - 1 > 2x + 9$ 8. $6x + 1 \geq 4x - 3$ 7. _____

                                                                        8. _____

9. $3x - 2 > 7$ 10. $2x + 3 < 15$ 9. _____

                                                                        10. _____

11. $4x - 3 \leq 9$ 12. $5x + 1 \leq -9$ 11. _____

                                                                        12. _____

13. $7x + 2 > 4x - 7$ 14. $6x + 5 < 4x - 5$ 13. _____

                                                                        14. _____

15. $5x + 7 > 4x - 2$ 16. $9x + 6 < 2x - 8$ 15. _____

                                                                        16. _____

17. $7x + 3 \geq 3x + 19$ 18. $6x - 1 < 2x + 7$ 17. _____

                                                                        18. _____

19. $7 - 6x \geq 31$ 20. $2 - 7x \leq 23$ 19. _____

                                                                        20. _____

21. $-4 - 3x > -7$ 22. $-3 - x < 5$ 21. _____

                                                                        22. _____

23. $7x - 3 < 5x - 11$ 24. $5x + 4 \geq x - 16$ 23. _____

                                                                        24. _____

Name   Score

Solve. Write the solution in set builder notation.

1. $2x < 8$ and $x + 3 > 1$

2. $x - 3 \le 5$ and $3x > -15$

3. $x + 1 \ge 7$ or $3x < 9$

4. $4x < 16$ or $x - 2 > 7$

5. $-2x > -12$ and $-4x < 12$

6. $\frac{1}{3}x > -1$ and $4x < 12$

7. $x + 3 \ge 6$ and $2x \ge 8$

8. $2x < -10$ and $x - 1 < 3$

9. $-3x > 9$ and $x + 2 > 5$

10. $3x + 1 < 7$ and $3x + 5 > -1$

11. $6x + 5 < 11$ or $3x - 1 > 8$

12. $5x - 3 < -18$ or $6x - 1 > 17$

Solve. Write the solution in interval notation.

13. $-2 < 3x + 7 < 10$

14. $-5 < 2x + 1 < 9$

15. $0 < 2x - 8 < 2$

16. $-2 < 3x + 7 < 4$

17. $3x - 2 > 7$ or $3x - 4 \le -10$

18. $5x - 7 > 9$ or $4x - 7 < -11$

19. $2x - 5 \ge 3$ and $3x - 2 > 10$

20. $5x - 2 < 3$ or $6x - 5 < 7$

21. $-7 \le 6x + 17 \le 35$

22. $2 \le 5x - 13 \le 12$

23. $5 - 8x \le 21$ and $4 - 5x > -1$

24. $7 - x \ge 9$ and $7 - 2x < 5$

1. _____
2. _____
3. _____
4. _____
5. _____
6. _____
7. _____
8. _____
9. _____
10. _____
11. _____
12. _____
13. _____
14. _____
15. _____
16. _____
17. _____
18. _____
19. _____
20. _____
21. _____
22. _____
23. _____
24. _____

21

Name   Score

Solve.

1. The length of a rectangle is 3 ft more than five times the width. Express as an integer the maximum width of the rectangle when the perimeter is less than 54 ft.

2. Three times the difference between a number and five is less than or equal to eight times the sum of the number and 10. Find the smallest number that will satisfy the inequality.

1. _____

2. _____

3. Suppose Lemon Rental Cars rents sedans for $39 per day with unlimited mileage, and Klunker Rentals offers sedans for $24.99 per day but charges $0.14 for each mile beyond 150 miles driven per day. You want to rent a car for a 10-day vacation. How many miles can you drive during the vacation if Klunker Rentals is to be less expensive than Lemon?

4. The temperature range for a week was between 23°F and 68°F. Find the temperature range in Celsius degrees.
$$C = \frac{5(F - 32)}{9}$$

3. _____

4. _____

5. Heather Malen earns $1500 per month plus 8% commission on the amount of sales. Heather's goal is to earn a minimum of $4060 per month. What amount of sales will enable Heather to earn $4060 or more per month?

6. Strong National Bank offers two different checking accounts. Plan A charges $2.50 per month and $0.25 per check after the first 10 checks. Plan B charges $7.50 per month with unlimited check writing. How many checks can be written per month if plan A is less expensive than plan B?

5. _____

6. _____

7. An average of 80 to 89 in an algebra class receives a B grade. A student has scores of 86, 95, 87, and 79 on four tests. Find the range of scores on the fifth test that will give the student a B for the course.

8. Find three consecutive odd integers whose sum is between 78 and 99.

7. _____

8. _____

Name   Score

Solve.

1. $|x| = 6$ **2.** $|a| = 3$ **3.** $|b| = 5$ **1.** _____

                                                                                 2. _____

                                                                                 3. _____

4. $|c| = 11$ **5.** $|-y| = 7$ **6.** $|-t| = 2$ **4.** _____

                                                                                 5. _____

                                                                                 6. _____

7. $|-a| = 9$ **8.** $|-x| = 1$ **9.** $|x| = -3$ **7.** _____

                                                                                 8. _____

                                                                                 9. _____

10. $|-x| = -5$ **11.** $|5x - 10| = 0$ **12.** $|2x - 6| = 8$ **10.** _____

                                                                                 11. _____

                                                                                 12. _____

13. $|4 - 2x| = 8$ **14.** $|a - 6| = 0$ **15.** $|3x - 2| = 0$ **13.** _____

                                                                                 14. _____

                                                                                 15. _____

16. $|3 - 2x| = 5$ **17.** $|2 - 3x| = 7$ **18.** $|2a + 7| = 1$ **16.** _____

                                                                                 17. _____

                                                                                 18. _____

19. $|5x - 1| - 1 = 3$ **20.** $|4x - 3| + 2 = 5$ **21.** $|3x + 2| - 1 = 3$ **19.** _____

                                                                                 20. _____

                                                                                 21. _____

22. $|4x + 3| - 1 = 8$ **23.** $|5 - 3x| + 4 = 1$ **24.** $3 - |3x + 1| = 3$ **22.** _____

                                                                                 23. _____

                                                                                 24. _____

25. $4 + |2x - 1| = 7$ **26.** $7 - |3x - 2| = 4$ **27.** $3 - |2x - 5| = 3$ **25.** _____

                                                                                 26. _____

                                                                                 27. _____

23

Name   Score

Solve.

1. $|x| > 2$ 2. $|x| < 4$ 1. _____

                                                                             2. _____

3. $|3 - x| \geq 1$ 4. $|2 - x| \geq 5$ 3. _____

                                                                             4. _____

5. $|2x - 1| < 7$ 6. $|x + 7| \geq 3$ 5. _____

                                                                             6. _____

7. $|x - 3| < 2$ 8. $|x - 4| \leq 7$ 7. _____

                                                                             8. _____

9. $|3x + 5| > 11$ 10. $|5x - 1| > 14$ 9. _____

                                                                             10. _____

11. $|2x - 3| \leq -3$ 12. $|4 - 3x| \leq 1$ 11. _____

                                                                             12. _____

13. $|3 - 4x| > 11$ 14. $|3x - 2| < 10$ 13. _____

                                                                             14. _____

15. $|18 - 2x| \leq 0$ 16. $|5x - 3| < 12$ 15. _____

                                                                             16. _____

17. $|3 - 2x| > 7$ 18. $|4x - 1| < 3$ 17. _____

                                                                             18. _____

Name _____   Score _____

Solve.

1. A doctor has prescribed 4 cc of medicine for a patient. The tolerance is 0.06 cc. Find the lower and upper limits of the amount of medication to be given.

2. The power of an electrical system is 105 volts plus or minus 20 volts. Find the lower and upper limits of voltage on which the system will run.

1. _____

2. _____

3. The power of a hand-held dryer is 115 volts plus or minus 17.5 volts. Find the lower and upper limits of voltage on which the dryer will run.

4. A doctor has prescribed 2.5 cc of medicine for a patient. The tolerance is 0.03 cc. Find the lower and upper limits of the amount of medicine to be given.

3. _____

4. _____

5. Find the lower and upper limits of a 28,000-ohm resistor with a 2% tolerance.

6. Find the lower and upper limits of a 16,000-ohm resistor with an 8% tolerance.

5. _____

6. _____

7. Find the lower and upper limits of a 72-ohm resistor with a 5% tolerance.

8. A small lamp is 6.25 volts plus or minus 0.5 volts. Find the lower and upper limits of voltage on which it will work.

7. _____

8. _____

Name   Score

Find the distance to the nearest hundredth between the given points.

1. $P_1(1, 4)$ and $P_2(6, 2)$ 2. $P_1(3, -2)$ and $P_2(-1, 1)$ 1. _____

                                                                                          2. _____

3. $P_1(-2, 5)$ and $P_2(-1, -1)$ 4. $P_1(-4, -1)$ and $P_2(-2, 3)$ 3. _____

                                                                                          4. _____

Find the coordinates of the midpoint of the line segment connecting the points.

5. $P_1(-5, -3)$ and $P_2(3, 3)$ 6. $P_1(4, 2)$ and $P_2(1, 1)$ 5. _____

                                                                                          6. _____

7. $P_1(2, 5)$ and $P_2(-2, 1)$ 8. $P_1(2, -3)$ and $P_2(-1, 4)$ 7. _____

                                                                                          8. _____

9. $P_1(-1, 3)$ and $P_2(3, -1)$ 10. $P_1(2, -2)$ and $P_2(1, -1)$ 9. _____

                                                                                          10. _____

Name   Score

Solve.

1. Graph the ordered pairs (2, 1) and (1, 3). Draw a line between the two points.

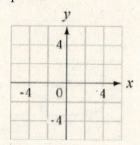

2. Graph the ordered pairs (−1, 2) and (2, −1). Draw a line between the two points.

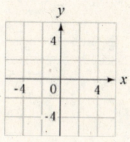

3. Graph the ordered pairs (−2, −3) and (3, 1). Draw a line between the two points.

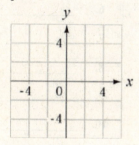

4. Graph the ordered pairs (0, −3) and (−2, 0). Draw a line between the two points.

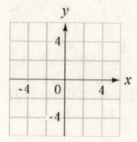

5. Find the coordinates of each of the points.

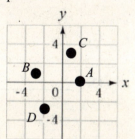

6. Find the coordinates of each of the points.

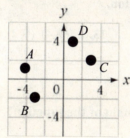

5. _____

6. _____

7. Draw a line through all points with an abscissa of −2.

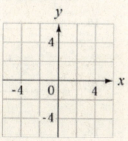

8. Draw a line through all points with an ordinate of 3.

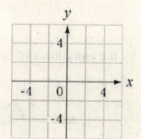

Name _____   Score _____

State whether the relation is a function.

1.   {(1, 1), (2, 3), (4, 3), (5, 7)}   2.   {(−3, −1), (−1, −2), (1, 5), (2, 6), (1, −7)}   1. _____

                                                                                2. _____

Given $f(x) = 2x + 1$**, evaluate:**

3. $f(-1)$ 4. $f(2)$ 5. $f(0)$ 3. _____

                                                                4. _____

                                                                5. _____

Given $g(x) = x^2 + x - 1$**, evaluate:**

6. $g(0)$ 7. $g(1)$ 8. $g(-1)$ 6. _____

                                                                7. _____

                                                                8. _____

9. $g(2)$ 10. $g(-2)$ 11. $g(t)$ 9. _____

                                                                10. _____

                                                                11. _____

Find the domain and range of the function.

12.   {(1, 1), (2, 3), (4, 3), (5, 7)}   13.   {(1, 4), (6, 2), (3, 1), (5, 4)}   12. _____

                                                                13. _____

What values are excluded from the domain of the function?

14. $f(x) = \dfrac{3x + 5}{6}$ 15. $g(x) = \dfrac{1}{x + 5}$ 16. $h(x) = \dfrac{4 - x}{3 - x}$ 14. _____

                                                                15. _____

                                                                16. _____

Find the range of the function defined by the equation and the given domain.

17. $f(x) = 4x + 3$ 18. $g(x) = \dfrac{3}{4 - x}$ 17. _____
      domain = {−1, 1, 3}   domain = {−1, 0, 3}

                                                                18. _____

Name   Score

Graph.

1. $y = 2x - 1$

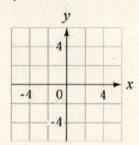

2. $y = 3x - 3$

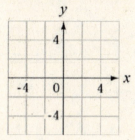

3. $y = \dfrac{1}{3}x$

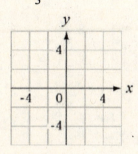

4. $y = \dfrac{4}{3}x$

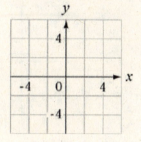

5. $y = -\dfrac{1}{3}x + 2$

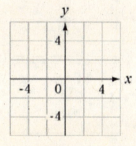

6. $y = \dfrac{1}{4}x - 1$

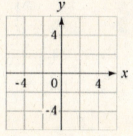

7. $y = -\dfrac{1}{2}x + 3$

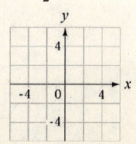

8. $y = -\dfrac{3}{2}x + 1$

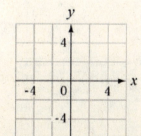

Name   Score

Graph.

1. $x - 2y = -4$

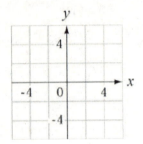

2. $4x - 2y = 8$

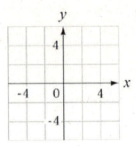

3. $3x - 2y = 6$

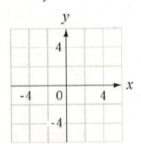

4. $y = 2$

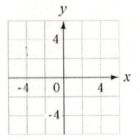

5. $y = -\dfrac{1}{3}x$

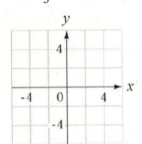

6. $x + 4y = 8$

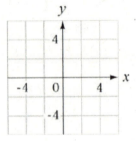

7. $2x - 5y = -10$

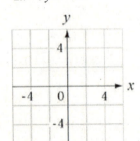

8. $3x + y = -3$

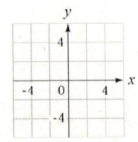

Name   Score

Find the *x*– and *y*–intercepts and graph.

1. $y = \dfrac{2}{5}x - 2$

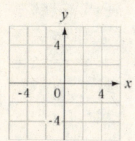

2. $3x - y = 6$

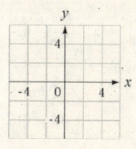

3. $2x + 3y = 4$

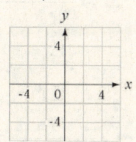

4. $y = -2x + 2$

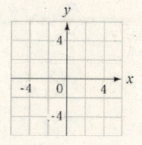

5. $y = -\dfrac{1}{3}x - 2$

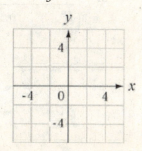

6. $3x + y = -3$

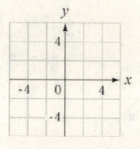

7. $y = 4x$

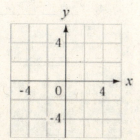

8. $2x - y = 5$

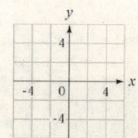

Name   Score

Solve.

1. Loren receives $10 per hour as a
    mathematics tutor. The equation that
    describes her wages is $w = 10t$, where t is
    the number of hours she spends tutoring.
    Graph this equation for $0 \leq t \leq 20$. The
    ordered pair (16, 160) is on the graph. Write
    a sentence that describes the meaning of
    this ordered pair.

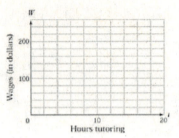

1. _____

2. A roller coaster has a maximum
    speed of 99 ft/s. The equation that
    describes the total number of feet
    traveled by the roller coaster in t seconds
    at this speed is given by $D = 99t$. Graph
    this equation for $0 \leq t \leq 10$. The point
    (6, 594) is on this graph. Write a
    sentence that describes the meaning of
    this ordered pair.

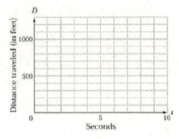

2. _____

3. The monthly cost for receiving messages
    from a telephone answering service
    is $7.00 plus $0.25 per message. The
    equation that describes the cost is
    $C = 0.25n + 7.00$, where n is the number
    of messages received. Graph the equation
    for $0 \leq n \leq 40$. The point (36, 16) is on
    the graph. Write a sentence that describes
    the meaning of this ordered pair.

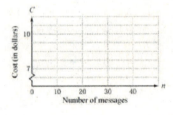

3. _____

4. The cost of manufacturing snow boards
    is $4500 for startup and $75 per board
    manufactured. The equation that describes
    the cost of manufacturing n boards is
    $C = 75n + 4500$. Graph this equation for
    $0 \leq n \leq 100$. The point (60, 9000) is on the
    graph. Write a sentence that describes the
    meaning of this ordered pair.

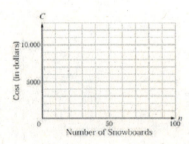

4. _____

Name   Score

Find the slope of the line containing the points.

1. $P_1(1, -3), P_2(5, -1)$ **2.** $P_1(-6, 7), P_2(2, 3)$ **3.** $P_1(0, -3), P_2(-1, 0)$ **1.** _____

2. _____

3. _____

4. $P_1(1, 2), P_2(3, 4)$ **5.** $P_1(3, 2), P_2(2, 1)$ **6.** $P_1(-2, 1), P_2(-6, 3)$ **4.** _____

5. _____

6. _____

7. $P_1(2, 3), P_2(3, 2)$ **8.** $P_1(4, -5), P_2(1, -4)$ **9.** $P_1(-1, -6), P_2(-4, -1)$ **7.** _____

8. _____

9. _____

10. $P_1(-4, -5), P_2(-2, 0)$ **11.** $P_1(4, 3), P_2(2, 1)$ **12.** $P_1(1, 2), P_2(4, 8)$ **10.** _____

11. _____

12. _____

13. $P_1(-4, 4), P_2(2, -2)$ **14.** $P_1(-3, 4), P_2(-1, -6)$ **15.** $P_1(0, -4), P_2(3, -10)$ **13.** _____

14. _____

15. _____

16. $P_1(-1, 3), P_2(2, 4)$ **17.** $P_1(2, -2), P_2(0, 4)$ **18.** $P_1(-1, 2), P_2(-3, 4)$ **16.** _____

17. _____

18. _____

19. $P_1(1, 2), P_2(-1, 2)$ **20.** $P_1(1, -2), P_2(5, -1)$ **21.** $P_1(2, 0), P_2(1, -1)$ **19.** _____

20. _____

21. _____

Name   Score

Graph by using the slope and the *y*–intercept.

1. $y = \frac{1}{2}x + 3$

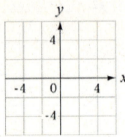

2. $y = \frac{3}{2}x + 2$

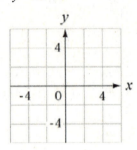

3. $y = -\frac{4}{3}x$

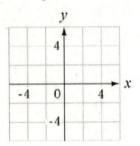

4. $y = 2x - 4$

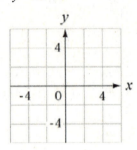

5. $3x - y = 3$

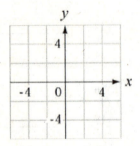

6. $2x + y = 4$

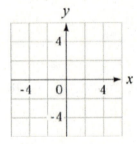

7. Graph the line that passes through point (1, 2) and has slope $\frac{1}{3}$.

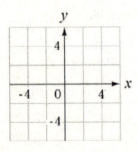

8. Graph the line that passes through point (−2, 1) and has slope −1.

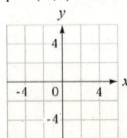

Name   Score

Find the equation of the line that contains the given point and has the given slope.

1.   Point $(0, 5)$. $m = -3$ **2.**   Point $(-2, 0)$. $m = 4$ **1.** _____

                                                                                        2. _____

3.   Point $(1, 2)$. $m = \frac{1}{3}$ **4.**   Point $(3, -2)$. $m = -\frac{1}{3}$ **3.** _____

                                                                                        4. _____

5.   Point $(-3, 2)$. $m = -1$ **6.**   Point $(-3, 2)$. $m = \frac{1}{2}$ **5.** _____

                                                                                        6. _____

7.   Point $(-1, 1)$. $m = \frac{2}{3}$ **8.**   Point $(0, 3)$. $m = -\frac{1}{3}$ **7.** _____

                                                                                        8. _____

9.   Point $(-1, 5)$. $m = -2$ **10.** Point $(-1, 3)$. $m = -\frac{2}{5}$ **9.** _____

                                                                                        10. _____

11. Point $(1, -2)$. $m = \frac{3}{4}$ **12.** Point $(-1, -3)$. $m = \frac{1}{3}$ **11.** _____

                                                                                        12. _____

13. Point $(-1, -1)$. $m = -\frac{1}{2}$ **14.** Point $(0, 0)$. $m = -\frac{1}{4}$ **13.** _____

                                                                                        14. _____

15. Point $(3, -2)$. $m = 2$ **16.** Point $(3, -4)$. $m = -2$ **15.** _____

                                                                                        16. _____

17. Point $(2, 4)$. $m = -\frac{1}{3}$ **18.** Point $(4, 1)$. $m = -\frac{3}{5}$ **17.** _____

                                                                                        18. _____

Name   Score

Find the equation of the line that contains the given two points.

1. $P_1(1, 4)$, $P_2(2, 3)$ 2. $P_1(1, 3)$, $P_2(0, 4)$ 1. _____

                                                                          2. _____

3. $P_1(1, 2)$, $P_2(4, 4)$ 4. $P_1(3, 1)$, $P_2(5, 2)$ 3. _____

                                                                          4. _____

5. $P_1(-2, 2)$, $P_2(1, 3)$ 6. $P_1(-2, -2)$, $P_2(1, 1)$ 5. _____

                                                                          6. _____

7. $P_1(-2, -3)$, $P_2(2, 3)$ 8. $P_1(-2, 1)$, $P_2(1, 5)$ 7. _____

                                                                          8. _____

9. $P_1(1, 0)$, $P_2(0, -2)$ 10. $P_1(0, 3)$, $P_2(-1, 0)$ 9. _____

                                                                          10. _____

11. $P_1(0, 0)$, $P_2(2, 1)$ 12. $P_1(1, -3)$, $P_2(0, 0)$ 11. _____

                                                                          12. _____

13. $P_1(1, 0)$, $P_2(-2, 2)$ 14. $P_1(2, -4)$, $P_2(-1, 0)$ 13. _____

                                                                          14. _____

15. $P_1(3, -2)$, $P_2(-2, 4)$ 16. $P_1(4, -5)$, $P_2(-3, 1)$ 15. _____

                                                                          16. _____

17. $P_1(-1, 3)$, $P_2(2, 0)$ 18. $P_1(2, -2)$, $P_2(3, -2)$ 17. _____

                                                                          18. _____

Name   Score

Solve.

1. A building contractor estimates that the cost to build a home is $35,000 plus $95 for each square foot of floor space in the house.
   a. Determine a linear function that will give the cost of building a house that contains a give number of square feet.
   b. Use this model to determine the cost to build a house containing 1500 sq. ft.

2. The gas tank of a certain car contains 13 gal when the driver of the car begins a trip. Each mile driven by the driver decreases the amount of gas in the tank by 0.025 gal.
   a. Write a linear function for the number of gallons of gas in the tank in terms of the number of miles driven.
   b. Use your equation to find the number of gallons in the tank after 180 miles are driven.

1. a. _____

   b. _____

2. a. _____

   b. _____

3. A manufacturer of minivans determined that 75,000 cars per month can be sold at a price of $24,000. At a price of $23,500 the number of minivan sold per month would increase to 80,000.
   a. Determine a linear function that will predict the number of cars that would be sold each month at a given price.
   b. Use this model to predict the number of cars that would be sold at a price of $23,250.

4. An account executive receives a base salary plus a commission. On $25,000 in monthly sales, the account executive receives $2450. On $40,000 in monthly sales, the account executive receives $3200.
   a. Determine a linear function that will yield the compensation of the account executive for a given amount of monthly sales.
   b. Use this model to determine the account executive's compensation for $70,000 in monthly sales.

3. a. _____

   b. _____

4. a. _____

   b. _____

5. There are approximately 90 Calories in a 4 oz serving of cottage cheese and approximately 135 Calories in a 6 oz serving.
   a. Determine a linear function for the number of Calories in a serving of cottage cheese in terms of the size of the serving.
   b. Use your equation to estimate the number of Calories in a 5 oz serving.

6. A cellular phone company offers a plan for people who plan to use the phone only in emergencies. The plan costs the user $7.95 per month plus $0.69 per minute used.
   a. Write a linear function for the monthly cost in terms of the number of minutes used.
   b. Use your equation to find the cost of using the cellular phone for 9 minutes in one month.

5. a. _____

   b. _____

6. a. _____

   b. _____

7. At sea level, the boiling point of water is 100°C. At an altitude of 1.5 km, the boiling point of water is 94.75°C.
   a. Write a linear function for the boiling point of water in terms of the altitude above sea level.
   b. Use your equation to predict the boiling point of water at an altitude of 5 km above sea level.

8. Let f be a linear function. If $f(-3) = -1$ and $f(2) = -11$, find $f(x)$.

7. a. _____

   b. _____

8. _____

Name   Score

Solve.

1. Is the line $x = -1$ perpendicular to the line $y = 2$?

2. Is the line $y = \dfrac{1}{4}$ perpendicular to the line $y = -2$?

3. Is the line $y = 2x - 5$ parallel to the line $y = -\dfrac{1}{2}x - 5$?

4. Is the line $y = -\dfrac{1}{3}x + 2$ parallel to the line $y = -\dfrac{1}{3}x - 1$?

5. Is the line $y = \dfrac{3}{2}x - 1$ perpendicular to the line $y = -\dfrac{2}{3}x + 3$?

6. Are the lines $x + 2y = 1$ and $x + 2y = -3$ parallel?

7. Are the lines $x - 3y = 4$ and $3x + y = 6$ perpendicular?

8. Are the line $3x - 2y = 5$ and $3x + 2y = -3$ perpendicular?

9. Is the line that contains the points $(4, -5)$ and $(11, -10)$ perpendicular to the line that contains the points $(3, 7)$ and $(-2, 0)$?

10. Find the equation of the line containing the point $(-2, -5)$ and parallel to the line $x + 3y = 5$.

11. Find the equation of the line containing the point $(3, -1)$ and perpendicular to the line $y = -4x + 7$.

12. Find the equation of the line containing the point $(4, 3)$ and parallel to the line $5x - 3y = 8$.

13. Find the equation of the line containing the point $(3, -2)$ and perpendicular to the line $y = 2x$.

14. Find the equation of the line containing the point $(-1, 4)$ and perpendicular to the line $y = \dfrac{3}{4}x - 4$.

15. Find the equation of the line containing the point $(-2, -2)$ and parallel to the line $2x - 3y = 3$.

16. Find the equation of the line containing the point $(-1, 5)$ and parallel to the line $3x + y = -2$.

1. _____
2. _____
3. _____
4. _____
5. _____
6. _____
7. _____
8. _____
9. _____
10. _____
11. _____
12. _____
13. _____
14. _____
15. _____
16. _____

Name　　　　　　　　　　　　　　　　　　　　　　　　　　　　Score

Graph the solution set.

1.　　$x + y \geq 2$

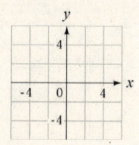

2.　　$x + 2y < 4$

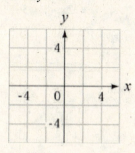

3.　　$x + 3y \leq 6$

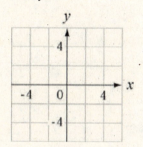

4.　　$3x + 2y \leq 6$

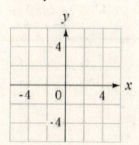

5.　　$5x - 2y < 10$

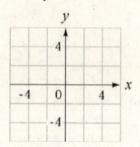

6.　　$3x - 4y \leq 12$

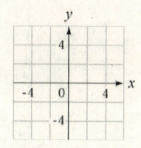

7.　　$y - 3 > 0$

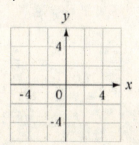

8.　　$x - 1 \geq 0$

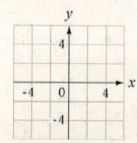

1. _____

2. _____

3. _____

4. _____

5. _____

6. _____

7. _____

8. _____

Name   Score

Solve by graphing.

1. $x + y = -1$
    $x + 2y = 1$

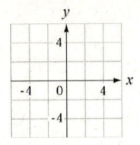

2. $x + y = 1$
    $2x - y = 5$

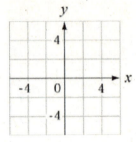

3. $x - y = 1$
    $2x - y = -1$

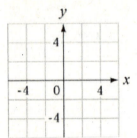

4. $x - 2y = 1$
    $x + y = 4$

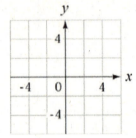

5. $2x - 3y = 10$
    $y = -2$

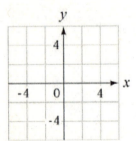

6. $x = 3$
    $2x - y = 4$

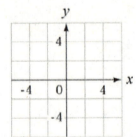

7. $x = 2$
    $y = 1$

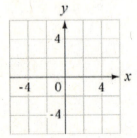

8. $x + 1 = 0$
    $y - 3 = 0$

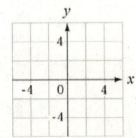

Name _____ Score

Solve by substitution.

1. $2x - y = 4$
    $x = 3$

2. $y = -1$
    $3x + 2y = 10$

3. $y = x - 2$
    $2x - y = 5$

4. $y = -x + 2$
    $3x - y = -6$

5. $x = y + 2$
    $x - 2y = -1$

6. $x = 3y - 1$
    $3x - y = 5$

7. $3x - 2y = 7$
    $y = 2x - 4$

8. $4x + y = -5$
    $y = 3x + 9$

9. $2x - 5y = -1$
    $y = 7 - 3x$

10. $5x - 3y = -7$
    $x = y - 1$

11. $3x + 2y = 3$
    $y = 1 - 2x$

12. $5x + 2y = 4$
    $y = x - 5$

13. $3x + y = 1$
    $2x + 2y = 6$

14. $x + 3y = -5$
    $2x - y = -3$

15. $x - 2y = 2$
    $2x - 5y = 3$

16. $4x - 2y = 5$
    $-x + 4y = -3$

17. $2x + 5y = -6$
    $-x + 3y = -8$

18. $3x + 2y = 5$
    $2x + y = 3$

19. $x = 3y + 1$
    $x = 2y - 1$

20. $x = 3y + 2$
    $x = -y - 6$

21. $y = 4x - 1$
    $y = 4 - x$

22. $y = 4 - 3x$
    $y = 3 - 2x$

23. $-x + 2y = 8$
    $4x + y = 13$

24. $2x + y = 3$
    $3x - 2y = -20$

1. _____

2. _____

3. _____

4. _____

5. _____

6. _____

7. _____

8. _____

9. _____

10. _____

11. _____

12. _____

13. _____

14. _____

15. _____

16. _____

17. _____

18. _____

19. _____

20. _____

21. _____

22. _____

23. _____

24. _____

Name   Score

Solve.

1. A total of $10,000 is deposited into two simple interest accounts. On one account, the simple interest rate is 4.5%; on the second account the annual simple interest rate is 6%. How much should be invested in each account so that the total annual interest earned is $504?

2. An investment club invested part of $14,000 in a 10% tax-free annual simple interest account and the remainder in a 14.5% annual simple interest account. The amount of interest earned for one year was $1670. How much was invested in each account?

1. _____

2. _____

3. An investment of $6000 is made at an annual simple interest rate of 9.75%. How much additional money must be invested at an annual simple interest rate of 14% so that the total interest earned is 11% of the total investment?

4. An investment of $7500 is made at 10.4%. How much additional money must be invested at an annual simple interest rate of 14% so that the total interest earned is 12% of the total investment?

3. _____

4. _____

5. An investment of $12,000 is deposited into two simple interest accounts. On one account the annual simple interest rate is 11%. On the other, the annual simple interest rate is 13%. How much should be invested in each account so that each account earns the same interest?

6. An investment advisor deposits $38,000 into two simple interest accounts. On the tax-free account the annual simple interest rate is 9%; on the money market fund the annual simple interest rate is 15%. How much should be invested in each account so that each account earns the same interest?

5. _____

6. _____

7. An investment advisor invested $24,000 in two accounts. One investment earned 16% annual simple interest; the other investment lost 5%. The total earnings from both investments were $1740. Find the amount invested at 16%.

8. An investment advisor invested $18,000 in two accounts. One investment earned 13% annual simple interest; the other investment lost 4%. The total earnings from both investments were $980. Find the amount invested at 13%.

7. _____

8. _____

9. A total of $14,000 is invested in two simple interest accounts. On one account the annual simple interest rate is 12.5%. On the second account the annual simple interest rate is 10%. The total annual interest earned by the two accounts is $1600. How much is invested in the 10% account?

10. An investment advisor invested $20,000 into two simple interest accounts. One investment earned 14% annual simple interest; the other investment earned 12.8% annual simple interest. The amount of interest earned in one year was $2734. How much was invested in the 14% account?

9. _____

10. _____

Name   Score

Solve by the addition method.

1. $x - y = 3$
    $x + y = 5$

2. $x + y = 5$
    $3x - y = 7$

3. $x - y = 7$
    $3x + y = 9$

4. $x - 2y = 3$
    $x + 3y = -2$

5. $2x + y = 1$
    $x + 2y = 5$

6. $x - 4y = -10$
    $3x - 2y = -10$

7. $2x - 3y = 7$
    $3x + y = 16$

8. $2x - y = 1$
    $3x - 2y = -1$

9. $x + 3y = -2$
    $2x - y = 10$

10. $x + 2y = 12$
    $3x - 2y = -4$

11. $2x - 3y = 7$
    $4x - 6y = 14$

12. $2x - y = 4$
    $6x - 3y = 12$

13. $8x - y = 3$
    $4x + 2y = 4$

14. $3x + 7y = 1$
    $4x + 5y = -3$

15. $6x + 2y = 0$
    $2x - 5y = 17$

16. $2x + 11y = -1$
    $3x + 2y = 13$

17. $2x + 3y = -8$
    $5x - 6y = 7$

18. $2x - 5y = 19$
    $3x - 2y = 12$

19. $3x - 4y = -11$
    $2x - 3y = -8$

20. $4x - 11y = 38$
    $5x + 2y = 16$

21. $2x - 3y = -7$
    $3x + 2y = -4$

22. $4x + 3y = 2x + 5$
    $3x - 2y = x + 10$

23. $5x + y = 2x - 4$
    $4x - 2y = 3x + 7$

24. $\dfrac{3}{5}x - \dfrac{1}{3}y = 7$

    $\dfrac{1}{4}x - \dfrac{2}{5}y = -1$

1. _____

2. _____

3. _____

4. _____

5. _____

6. _____

7. _____

8. _____

9. _____

10. _____

11. _____

12. _____

13. _____

14. _____

15. _____

16. _____

17. _____

18. _____

19. _____

20. _____

21. _____

22. _____

23. _____

24. _____

Name　　　　　　　　　　　　　　　　　　　　　　　Score

Solve by the addition method.

1. $\begin{array}{l}2x+2y+z=1\\x-y+z=0\\x+y-z=-4\end{array}$

2. $\begin{array}{l}x+y-z=1\\2x+3y+z=3\\x-y-z=5\end{array}$

1. _____

2. _____

3. $\begin{array}{l}x+2y-z=8\\-x+2y+z=9\\x+y+z=-4\end{array}$

4. $\begin{array}{l}2x-y+z=-9\\3x+y-2x=-7\\x+4y+3z=16\end{array}$

3. _____

4. _____

5. $\begin{array}{l}2x+y+3z=8\\x+y+z=1\\4x-y-z=14\end{array}$

6. $\begin{array}{l}x+y-z=-2\\2x-y+z=-1\\x+2y-3z=-7\end{array}$

5. _____

6. _____

7. $\begin{array}{l}3x-y+z=4\\2x-3y+2z=6\\x-y+z=2\end{array}$

8. $\begin{array}{l}2x-y-3z=-4\\x+y+z=0\\x-y+4z=-9\end{array}$

7. _____

8. _____

9. $\begin{array}{l}4x-y+3z=35\\x-3y-2z=9\\3x+2y+6z=20\end{array}$

10. $\begin{array}{l}3x+4y-2z=-1\\5x+5y-3z=2\\6x+3y-2z=5\end{array}$

9. _____

10. _____

Name   Score

Evaluate the determinant.

1. $\begin{vmatrix} 4 & 1 \\ -2 & 2 \end{vmatrix}$

2. $\begin{vmatrix} 2 & 5 \\ 3 & 4 \end{vmatrix}$

3. $\begin{vmatrix} 6 & -9 \\ 2 & -1 \end{vmatrix}$

4. $\begin{vmatrix} 2 & 3 \\ 1 & -4 \end{vmatrix}$

5. $\begin{vmatrix} 4 & 3 \\ 5 & -1 \end{vmatrix}$

6. $\begin{vmatrix} 2 & 1 \\ -3 & -8 \end{vmatrix}$

7. $\begin{vmatrix} 5 & 1 \\ -4 & 6 \end{vmatrix}$

8. $\begin{vmatrix} -2 & 3 \\ 1 & -2 \end{vmatrix}$

9. $\begin{vmatrix} 5 & -3 \\ 4 & -2 \end{vmatrix}$

10. $\begin{vmatrix} 3 & 4 & -1 \\ 2 & -1 & 6 \\ 1 & 2 & 3 \end{vmatrix}$

11. $\begin{vmatrix} 5 & 1 & 5 \\ -3 & 1 & 2 \\ 1 & 2 & 4 \end{vmatrix}$

12. $\begin{vmatrix} 2 & 5 & -2 \\ 3 & -2 & 5 \\ -2 & -3 & 2 \end{vmatrix}$

13. $\begin{vmatrix} 1 & 4 & 2 \\ 2 & 3 & 1 \\ 6 & 2 & 4 \end{vmatrix}$

14. $\begin{vmatrix} 2 & 1 & 4 \\ 5 & 6 & 5 \\ -1 & 2 & -1 \end{vmatrix}$

15. $\begin{vmatrix} 1 & 3 & 4 \\ 1 & 3 & 2 \\ -1 & 5 & 2 \end{vmatrix}$

16. $\begin{vmatrix} 2 & -2 & 7 \\ 1 & -1 & 3 \\ 3 & 4 & 2 \end{vmatrix}$

17. $\begin{vmatrix} 1 & 5 & -2 \\ 2 & -3 & 3 \\ 1 & 3 & 3 \end{vmatrix}$

18. $\begin{vmatrix} 1 & 0 & 2 \\ 3 & -1 & 3 \\ 2 & 2 & 8 \end{vmatrix}$

1. _____

2. _____

3. _____

4. _____

5. _____

6. _____

7. _____

8. _____

9. _____

10. _____

11. _____

12. _____

13. _____

14. _____

15. _____

16. _____

17. _____

18. _____

Name   Score

Solve by using Cramer's Rule.

1. $2x + y = 10$
    $3x - y = -5$

2. $5x - 3y = 3$
    $2x + y = -1$

1. _____

2. _____

3. $2x - 3y = 5$
    $5x - 2y = 18$

4. $x - 3y = 6$
    $3x + 5y = 4$

3. _____

4. _____

5. $3x - 2y = -5$
    $8x - 3y = 3$

6. $2x + 5y = -3$
    $3x - 7y = 10$

5. _____

6. _____

7. $6x - 5y = 2$
    $9x + 10y = 38$

8. $2x + 5y = -31$
    $8x - 3y = -9$

7. _____

8. _____

9. $3x + 5y = 9$
    $9x - 2y = -7$

10. $5x + 3y = 6$
    $4x - 2y = -4$

9. _____

10. _____

11. $x + 3y - z = -2$
    $2x - y + z = 1$
    $x + 2y - z = -3$

12. $3x + y - z = 0$
    $2x + 2y - z = 1$
    $x + 3y + z = 12$

11. _____

12. _____

13. $x + 3y - z = -2$
    $2x + y - 2z = 1$
    $3x - y - 2z = 5$

14. $x + 3y + z = 5$
    $3x - y + 2z = 2$
    $2x + y - z = -4$

13. _____

14. _____

15. $5x - 3y + z = -10$
    $-3x + 4y - z = 10$
    $2x + 2y + 3z = 5$

16. $3x - 2y + 4z = 7$
    $5x - 3y - z = 12$
    $5x + 3y + 5z = 15$

15. _____

16. _____

Name   Score

Solve.

1. Flying with the wind, a small plane flew 615 mi in 3 h. Against the wind, the plane could fly only 405 mi in the same amount of time. Find the rate of the plane in calm air and the rate of the wind.

2. A jet plane flying with the wind went 2750 mi in 5 h. Against the wind, the plane could fly only 2450 mi in the same amount of time. Find the rate of the plane in calm air and the rate of the wind.

1. _____

2. _____

3. A cabin cruiser traveling with the current went 60 mi in 3 h. Against the current, it took 6 h to travel the same distance. Find the rate of the cabin cruiser in calm water and the rate of the current.

4. A motorboat traveling with the current went 72 mi in 4 h. Against the current, it took 6 h to travel the same distance. Find the rate of the boat in calm water and the rate of the current.

3. _____

4. _____

5. Flying with the wind, a pilot flew 800 mi between two cities in 4 h. The return trip against the wind took 5 h. Find the rate of the plane in calm air and the rate of the wind.

6. A corporate jet flying with the wind flew 1000 mi in 4 h. Flying against the wind, the plane required 5 h to travel the same distance. Find the rate of the plane in calm air.

5. _____

6. _____

7. A motorboat traveling with the current went 112 km in 4 h. Against the current, the boat could go only 80 km in the same amount of time. Find the rate of the boat in calm water and the rate of the current.

8. A rowing team rowing with the current traveled 45 mi in 3 h. Against the current, the team rowed 27 mi In 3 h. Find the rate of the rowing team in calm water and the rate of the current.

7. _____

8. _____

9. A plane flying with a tailwind flew 480 mi in 3 h. Against the wind, the plane required 4 h to fly the same distance. Find rate of the plane in calm air and the rate of the wind.

10. Flying with the wind, a plane flew 1080 mi in 3 h. Against the wind, the plane required 4 h to fly the same distance. Find the rate of the plane in calm air and the rate of the wind.

9. _____

10. _____

Name   Score

Solve.

1. A carpenter purchased 50 ft of redwood and 90 ft of pine for a total cost of $33. A second purchase, at the same prices, included 80 ft of redwood and 70 ft of pine for a total cost of $38. Find the cost per foot of redwood and of pine.

2. A merchant mixed 15 lb of a cinnamon tea with 10 lb of spice tea. The 25-pound mixture sells for $65. A second mixture included 14 lb of the cinnamon tea and 16 lb of the spice tea. The 30-pound mixture sells for $84. Find the cost per pound for the cinnamon and for the spice tea.

1. _____

2. _____

3. During one month, a small business used 900 units of electricity and 200 units of gas for a total cost of $218. The next month, 800 units of electricity and 250 units of gas were used for a total cost of $201. Find the cost per unit of gas.

4. A contractor buys 20 yd of nylon carpet and 30 yd of wool carpet for $860. A second purchase, at the same prices, includes 24 yd of nylon carpet and 32 yd of wool carpet for $984. Find the cost per yard of the wool carpet.

3. _____

4. _____

5. The total value of the quarters and nickels in a coin bank is $14.00. If the quarters were nickels and the nickels were quarters, the total value of the coins would be $22.00. Find the number of quarters in the bank.

6. A coin bank contains only quarters and dimes. The total value of the coins in the bank is $8.80. If the dimes were quarters and the quarters were dimes, the total value of the coins would be $7.30. Find the number of dimes in the bank.

5. _____

6. _____

7. A restaurant manager buys 120 lb of hamburger and 60 lb of steak for a total cost of $360. A second purchase, at the same prices, includes 200 lb of hamburger and 80 lb of steak. The total cost is $540. Find the cost of 1 lb of steak.

8. A sheet metal shop ordered 60 lb of tin and 30 lb of a zinc alloy for a total cost of $480. A second purchase, at the same prices, included 40 lb of tin and 70 lb of the zinc alloy. The total cost was $770. Find the total cost per pound of the tin and of the zinc alloy.

7. _____

8. _____

9. A citrus fruit grower purchased 30 orange trees and 35 grapefruit trees for $450. The next week, at the same prices, the grower bought 40 orange trees and 40 grapefruit trees for $560. Find the cost of an orange tree and of a grapefruit tree.

10. A store owner purchased thirty 75-watt light bulbs and 40 fluorescent light bulbs for a total cost of $95. A second purchase, at the same prices, included forty 75-watt light bulbs and 20 fluorescent lights for a total cost of $60. Find the cost of a 75-watt bulb and of a fluorescent light.

9. _____

10. _____

Name   Score

Graph the solution set.

1. $x - y \geq -2$
   $x + y \geq 3$

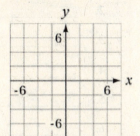

2. $2x - y \leq -3$
   $x + y \geq 2$

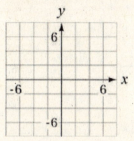

3. $x - 3y \leq 6$
   $2x + y \leq 4$

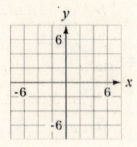

4. $x + 2y \leq 4$
   $x - y \leq 3$
   $y \geq 0$

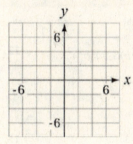

5. $x - 2y \geq 1$
   $2x + y \geq 2$

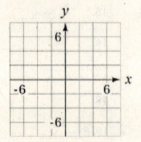

6. $x - 3y \leq 0$
   $2x - y \geq -3$

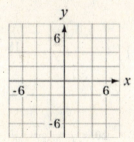

7. $2x - y \leq 0$
   $x - 2y \geq -2$

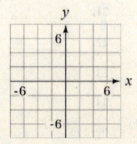

8. $x - 3y \geq -3$
   $x - 3y \leq 6$

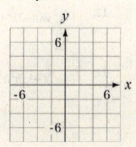

Name _____   Score _____

Simplify.

1. $(a^2 b)(a^2 b^3)$

2. $(-3x^2 y^2)(-4x^2 y^3)$

3. $(xy^3)^2$

4. $(x^3 y^2)^3$

5. $(-3a^2 b)^4$

6. $(-4x^2 y^3)^3$

7. $(2^2 a^4 b^2)^3$

8. $(x^2 y)(xy)^3$

9. $(xy)(x^3 y)^2$

10. $(5ab)^2 (-2a^2 bc^2)^3$

11. $(-4a^2 b)(-5a^3 b^4)^2$

12. $x^{3n} \cdot x^{3n}$

13. $x^n \cdot x^{n-2}$

14. $x^{4n} \cdot x^{3n+2}$

15. $x^{4n} \cdot x^{n-4}$

16. $\left(a^n\right)^{3n}$

17. $\left(a^{n-2}\right)^{3n}$

18. $\left(x^{3n-1}\right)^2$

19. $\left(x^{4n+1}\right)^4$

20. $\left(a^{3n-1}\right)^n$

21. $(-3x^3 y^2 z)(6xy^2 z^3)$

22. $(4x^2 yz)(-3x^2 y^2 z)(-5x^4 z^2)$

23. $(-4xy^2 z)^2 (-5x^2 yz^3)(2x^3 yz)$

24. $(-a^3)(-3ab^2 c^3)(4abc)$

1.	_____
2.	_____
3.	_____
4.	_____
5.	_____
6.	_____
7.	_____
8.	_____
9.	_____
10.	_____
11.	_____
12.	_____
13.	_____
14.	_____
15.	_____
16.	_____
17.	_____
18.	_____
19.	_____
20.	_____
21.	_____
22.	_____
23.	_____
24.	_____

Name   Score

Simplify.

1. 2^{-5}

2. $\dfrac{a^2}{3b^{-1}}$

3. xy^{-3}

4. $\dfrac{(3x)^0}{-3^2}$

5. $\dfrac{-2^{-3}}{(3x)^0}$

6. $a^{-2}\cdot a^3$

7. $\dfrac{x^{-2}}{x^3}$

8. $\dfrac{x^{-5}}{x^{-6}}$

9. $\dfrac{2^{-3}a^3}{2a^{-2}}$

10. $\dfrac{x^{-3}y^{-8}}{xy^{-5}}$

11. $\dfrac{x^3y^2}{x^{-2}y^{-1}}$

12. $\dfrac{a^{-2}b^{-4}}{a^3b^{-3}}$

13. $\dfrac{60a^9}{20a^6}$

14. $\dfrac{-4x^4y}{16x^2y^3}$

15. $\dfrac{a^3b^2c^6}{a^5bc^4}$

16. $\dfrac{a^4b^5c}{a^8b^3c^5}$

17. $\dfrac{\left(3xy^2\right)^2}{9x^2y^2}$

18. $\dfrac{\left(4x^2y^3\right)^2}{16x^3y^6}$

19. $\dfrac{-4a^2b}{\left(2ab^3\right)^3}$

20. $\left(\dfrac{-8a^3b^2}{12a^2b^3}\right)^3$

21. $\left(\dfrac{12x^2y^3z^4}{16xy^4z^2}\right)^4$

22. $\dfrac{\left(3xy^2\right)^2}{\left(6x^2y\right)^3}$

23. $\dfrac{\left(-2x^2y\right)^4}{\left(-4x^3y\right)^3}$

24. $\dfrac{a^{5n}}{a^{6n}}$

25. $\dfrac{a^{3n}}{-a^{6n}}$

26. $\dfrac{x^{3n-1}}{x^{n-2}}$

27. $\dfrac{a^{4n-3}b^{2n+2}}{a^{3n-1}b^{n+1}}$

1. _____ 2. _____ 3. _____ 4. _____ 5. _____ 6. _____ 7. _____ 8. _____ 9. _____ 10. _____ 11. _____ 12. _____ 13. _____ 14. _____ 15. _____ 16. _____ 17. _____ 18. _____ 19. _____ 20. _____ 21. _____ 22. _____ 23. _____ 24. _____ 25. _____ 26. _____ 27. _____

Name _____ Score _____

Write in scientific notation.

1. 0.0000356   **2.** 320,000   **3.** 0.00000037

4. 0.0000000051   **5.** 60,000,000,000   **6.** 46,000,000

Write in decimal notation.

7. 2.14×10^{-6} **8.** 5.3×10^{-9} **9.** 6.7×10^{9}

10. 5.23×10^{6} **11.** 4.5×10^{-3} **12.** 5.24×10^{7}

Simplify.

13. $\left(4 \times 10^{-10}\right)\left(7 \times 10^{7}\right)$ **14.** $\left(7.8 \times 10^{-4}\right)\left(2.1 \times 10^{-5}\right)$

15. $\left(5 \times 10^{-11}\right)\left(7 \times 10^{15}\right)$ **16.** $\left(6.7 \times 10^{-5}\right)\left(3.2 \times 10^{-3}\right)$

17. $(0.000047)(670,000,000)$ **18.** $(360,000)(0.00000062)$

19. $\dfrac{8 \times 10^{-2}}{2 \times 10^{7}}$ **20.** $\dfrac{2.8 \times 10^{5}}{4 \times 10^{-3}}$ **21.** $\dfrac{0.00072}{40,000,000}$

22. $\dfrac{3.600}{0.0000012}$ **23.** $\dfrac{0.000048}{0.00000008}$ **24.** $\dfrac{0.000000275}{0.0000005}$

1. _____
2. _____
3. _____
4. _____
5. _____
6. _____
7. _____
8. _____
9. _____
10. _____
11. _____
12. _____
13. _____
14. _____
15. _____
16. _____
17. _____
18. _____
19. _____
20. _____
21. _____
22. _____
23. _____
24. _____

Name   Score

Solve. Write the answer in scientific notation.

1. How many kilometers does light travel in 13 h? The speed of light is 300,000 km/s.

2. A computer can do an arithmetic operation in 8×10^{-6} s. How many arithmetic operations can the computer perform in 1 min?

1. _____

2. _____

3. A computer can do an arithmetic operation in 7×10^{-7} s. How many arithmetic operations can the computer perform in 1 h?

4. A high-speed centrifuge makes 8×10^{7} revolutions each minute. Find the time in second for the centrifuge to make 1 revolution.

3. _____

4. _____

5. A high-speed centrifuge makes 8×10^{6} revolutions each minute. How many revolutions does it make in 2 h?

6. It takes 4.0×10^{-9} s for light to travel 60 ft. How long does it take for light to travel 1 mi?

5. _____

6. _____

7. It takes 4.0×10^{-9} s for light to travel 60 ft. How long does it take for light to travel 10 mi?

8. A space vehicle travels at an average velocity of 6.0×10^{5} miles per day. How far does it travel each week?

7. _____

8. _____

9. How many miles doe slight travel in 4 days? The speed of light is 186,000 mi/s.

10. How many miles doe slight travel in 30 days? The speed of light is 186,000 mi/s.

9. _____

10. _____

Name   Score

Evaluate.

1.   Given $P(x) = 4x^3 - 3x^2 + 6$,   **2.**   Given $P(x) = 4x^3 - 3x^2 + 6$,   **1.** _____
      evaluate $P(-3)$.   evaluate $P(-3)$.
                                                                                          2. _____

Indicate which define a polynomial function. For those that are polynomial functions: a. Identify the leading coefficient. b. Identify the constant term. c. State the degree.

3. $P(x) = 6x^4 - 2x + 7$ **4.** $R(x) = \dfrac{x^2 + 2x + 1}{x}$ **3. a.** _____

                                                                       b. _____

                                                                       c. _____

                                                                    4. a. _____

                                                                       b. _____

                                                                       c. _____

5. $f(x) = x^2 - 6x - \sqrt{7}$ **6.** $g(x) = \pi$ **5. a.** _____

                                                                       b. _____

                                                                       c. _____

                                                                    6. a. _____

                                                                       b. _____

                                                                       c. _____

Graph.

7. $f(x) = x^2 - 3x + 2$ **8.** $f(x) = 2x^2 + 1$ **9.** $f(x) = 2x^2 - x - 3$

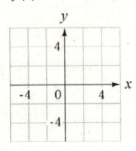   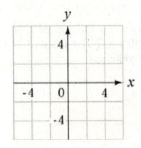   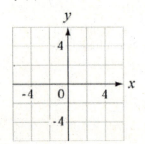

Name _____   Score _____

Simplify.

1. Given $P(x) = x^2 - 6$ and $R(x) = 4x - 2$, find $P(x) - R(x)$.

2. Given $P(x) = 3x^3 - 4x^2 - 6x$ and $R(x) = -6x^2 + 10x + 4$, find $P(x) - R(x)$.

1. _____

2. _____

Simplify. Use a vertical format.

3. $(x^2 - 4xy + 2y^2) + (3x^2 - 4y^2)$

4. $(2x^2 + 3y^2) + (-4x^2 + 3xy - 2y^2)$

3. _____

4. _____

5. $(2x^{2n} + 6x^n - 4) + (-x^{2n} - x^n + 7)$

6. $(x^{2n} - 2x^n - 2) + (4x^{2n} + 6n + 3)$

5. _____

6. _____

Simplify. Use a horizontal format.

7. $(4y^3 - 8y) + (y^2 - 7y + 1)$

8. $(-3y^2 - 5y - 11) + (6y^2 - 3y)$

7. _____

8. _____

9. $(3a^2 - 2a - 6) - (-3a^2 - a - 8)$

10. $(4x^4 - 2x^3 + x^2) + (4x^3 - 6x^2 + 3x)$

9. _____

10. _____

11. $(2x^4 - 3x + 2) + (4x^3 - 4x - 7)$

12. $(a^{2n} - 3a^n - 4) - (3a^{2n} - 2a^n + 7)$

11. _____

12. _____

13. $(x^{2n} - 2x^n - 5) - (3x^{2n} - x^n + 6)$

14. $(3x^2 - 2x - 8) - (-x^3 - 4x^2 + 5x) - (4x^3 - 2x^2 - 8x + 3)$

13. _____

14. _____

15. $(3a^3 - 3a^2b + 4ab^2 + 5b^3) - (5a^2b - 4ab^2 - 6b^3)$

15. _____

Name   Score

Simplify.

1. $3x(x-2)$ **2.** $4a(a+3)$ **3.** $2xy(4x-5y)$

1. _____

2. _____

3. _____

4. $-5ab(2a-3b)$ **5.** $x^n(x^{2n}-1)$ **6.** $x^{2n}(x^n-4)$

4. _____

5. _____

6. _____

7. $x^n(x^n-y^n)$ **8.** $x-3x(x-1)$ **9.** $4a+3a(2-a)$

7. _____

8. _____

9. _____

10. $-3a^2(2a^2-3a+4)$ **11.** $5b(2b^3-10b^2-4)$

10. _____

11. _____

12. $2b(4b^4-5b^2+7)$ **13.** $(3x^2-2x-6)(-3x^2)$

12. _____

13. _____

14. $(-4y^2-3y+4)(y^3)$ **15.** $(5b^3-4b^2-2)(-3b^2)$

14. _____

15. _____

16. $-4x^2(3-2x+2x^2+3x^3)$ **17.** $-3y(2-3y-4y^2+3y^3)$

16. _____

17. _____

18. $a^{n-1}(a^n-2a+3)$ **19.** $a^{n+3}(a^{n-1}+5a-2)$

18. _____

19. _____

20. $3y-2[y-3y(y-2)+5y]$ **21.** $3a^2-a[2-a(3-a-a^2)]$

20. _____

21. _____

Name _____

Score _____

Simplify.

1. $(x-1)(x+6)$

2. $(y+7)(y+2)$

3. $(y-3)(2y+1)$

4. $(3x-5)(2x-1)$

5. $(4a+5b)(3a+2b)$

6. $(2x-5y)(5x+3y)$

7. $(4x-7y)(5x-4y)$

8. $(xy-4)(xy+6)$

9. $(x^2-3)(x^2-5)$

10. $(3x^2-2y)(2x^2-y)$

11. $(x^2-y^2)(x^2+3y^2)$

12. $(x^n+3)(x^n-4)$

13. $(x^n-3)(x^n-4)$

14. $(2a^n-1)(3a^n+2)$

15. $(4x^n-3)(2x^n+1)$

16. $(3a^n-b^n)(2a^n+b^n)$

17. $(2a^n-1)(2a^n+1)$

18. $(x-1)(x^2-2x+5)$

19. $(x+3)(x^3-2x+2)$

20. $(a+1)(a^3-2a^2+5)$

21. $(x-4)(x^3+x-3)$

22. $(a-4)(a^2-a-2)$

23. $(y^2-2)(y^3-4y^2-2)$

24. $(x^2+1)(x^3+2x-3)$

1. _____
2. _____
3. _____
4. _____
5. _____
6. _____
7. _____
8. _____
9. _____
10. _____
11. _____
12. _____
13. _____
14. _____
15. _____
16. _____
17. _____
18. _____
19. _____
20. _____
21. _____
22. _____
23. _____
24. _____

Name

Score

Simplify.

1. $(a-3)(a+3)$

2. $(b-8)(b+8)$

3. $(4x-5y)(4x+5y)$

4. $(5x-3y)(5x+3y)$

5. $(xy+4)(xy-4)$

6. $(4a-3c)(4a+3c)$

7. $(2a+b)(2a-b)$

8. $(x^2+4)(x^2-4)$

9. $(x^2-7)(x^2+7)$

10. $(x^n+1)(x^n-1)$

11. $(a^n+b)(a^n-b)$

12. $(x-4y)(x+4y)$

13. $(x-6)^2$

14. $(4x-y)^2$

15. $(x^2-9)^2$

16. $(xy+3)^2$

17. $(xy-7)^2$

18. $\left(x^2-2y^2\right)^2$

19. $(x^2-1)^2$

20. $\left(x^2-y^2\right)^2$

21. $\left(3x^2+y^2\right)^2$

22. $\left(x^2-4\right)^2$

23. $\left(x^2-7\right)^2$

24. $\left(2x^2+5y^2\right)^2$

25. $\left(x^n+1\right)^2$

26. $\left(a^n+b^n\right)^2$

27. $\left(3x^n+4y^n\right)^2$

1. _____
2. _____
3. _____
4. _____
5. _____
6. _____
7. _____
8. _____
9. _____
10. _____
11. _____
12. _____
13. _____
14. _____
15. _____
16. _____
17. _____
18. _____
19. _____
20. _____
21. _____
22. _____
23. _____
24. _____
25. _____
26. _____
27. _____

Name _____   Score _____

Solve.

1. The length of a rectangle is $(4x - 5)$ ft. The width is $(x + 4)$ ft. Find the area of the rectangle in terms of the variable x.

2. The base of a triangle is $x - 5$ ft. The height is $2x + 4$ ft. Find the area of the triangle in terms of the variable x.

1. _____

2. _____

3. The length of the side of a square is $(3x + 2)$ ft. Find the area of the square in terms of the variable x.

4. The length of the side of a cube is $(x - 3)$ cm. Find the volume of the cube in terms of the variable x.

3. _____

4. _____

5. Find the area of the figure shown below. All dimensions given are in meters.

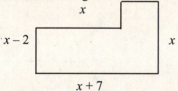

6. Find the area of the figure shown below. All dimensions given are in feet.

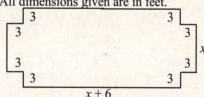

5. _____

6. _____

7. Find the volume of the figure shown below. All dimensions given are in inches.

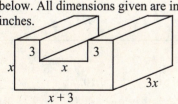

8. Find the volume of the figure shown below. All dimensions given are in centimeters.

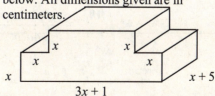

7. _____

8. _____

9. The radius of a circle is $(2x - 1)$ in. Find the area of the circle in terms of the variable x. Use 3.14 for π.

10. The radius of a circle is $(x + 2)$ in. Find the area of the circle in terms of the variable x. Use 3.14 for π.

9. _____

10. _____

Name _____   Score _____

Divide.

1. $\dfrac{3+9x}{3}$

2. $\dfrac{12x^2 - 4x}{4x}$

3. $\dfrac{12x^3 - 6x^2 + 18x}{6x}$

4. $\dfrac{2x^3y^3 - 18x^2y^2 + 6xy}{2xy}$

5. $\dfrac{-9a^4 - 18a^3 + 27a^2}{-9a^2}$

6. $\dfrac{4x^3 - 12x^2}{4x^2}$

7. $\dfrac{4x^2 + 6x}{2x}$

8. $\dfrac{25x^4 - 10x^3 + 5x^2}{-5x^2}$

9. $\dfrac{16y^6 + 20y^4 - 8y^3}{4y^2}$

10. $\dfrac{12a^6b^3 - 4a^5b^5}{-4a^4b^2}$

11. $\dfrac{7b^4 - 28b^2}{7b}$

12. $\dfrac{26a^4 + 52a^3}{-13a^2}$

1. _____

2. _____

3. _____

4. _____

5. _____

6. _____

7. _____

8. _____

9. _____

10. _____

11. _____

12. _____

Name _____ Score _____

Divide by using long division.

1. $(x^2 - 2x - 35) \div (x + 5)$

2. $(x^2 - 12x + 32) \div (x - 4)$

1. _____

2. _____

3. $(x^2 - x - 90) \div (x - 10)$

4. $(x^2 + 8x - 48) \div (x - 4)$

3. _____

4. _____

5. $(x^3 + 3x^2 - 5) \div (x - 2)$

6. $(6x^2 + x - 6) \div (3x + 2)$

5. _____

6. _____

7. $(12x^2 + 10x - 5) \div (2x + 1)$

8. $(15x^2 - x + 3) \div (3x + 1)$

7. _____

8. _____

9. $(6x^4 - 7x^2 - 8) \div (3x^2 - 2)$

10. $(27x^3 + 8) \div (3x + 2)$

9. _____

10. _____

11. $\dfrac{4x^3 - 6x^2 - 20x + 7}{2x - 1}$

12. $\dfrac{6x^3 - 7x^2 - 9x - 2}{3x + 1}$

11. _____

12. _____

13. $\dfrac{x^3 - 4x^2 + x + 6}{x - 2}$

14. $\dfrac{4x^3 - 8x^2 + 7x - 2}{2x - 1}$

13. _____

14. _____

15. $\dfrac{x^3 - 6x^2 + 8x - 12}{x - 4}$

16. $\dfrac{2x^3 - 7x^2 + 15x - 6}{2x - 1}$

15. _____

16. _____

17. $\dfrac{3x^4 - 20x^3 + 11x^2 + 8x - 12}{x - 6}$

18. $\dfrac{2x^4 + 5x^3 - 4x^2 - 2x + 3}{x + 3}$

17. _____

18. _____

61

Name   Score

Divide by using synthetic division.

1. $(2x^2 - x - 10) \div (x + 2)$ 2. $(4x^2 + 10x - 5) \div (x + 3)$ 1. _____

                                                                   2. _____

3. $(2x^2 - 17x + 35) \div (x - 5)$ 4. $(3x^2 - 7x - 15) \div (x - 4)$ 3. _____

                                                                   4. _____

5. $(4x^2 - 3) \div (x - 1)$ 6. $(4x^2 - 8) \div (x + 3)$ 5. _____

                                                                   6. _____

7. $(2x^3 + 3x^2 + 5x + 14) \div (x + 2)$ 8. $(3x^3 + 2x^2 - 8x + 3) \div (x - 1)$ 7. _____

                                                                   8. _____

9. $(3x^3 + x - 15) \div (x - 3)$ 10. $(x^3 - 4x^2 + 10) \div (x + 2)$ 9. _____

                                                                   10. _____

11. $(4x^3 + 15x^2 - 6x - 8) \div (x + 4)$ 12. $(3x^3 + 8x^2 - 6x + 2) \div (x - 1)$ 11. _____

                                                                   12. _____

13. $(2x^3 - 5x^2 - 6x + 10) \div (x - 5)$ 14. $(4x^3 + 9x^2 - 8x + 3) \div (x + 3)$ 13. _____

                                                                   14. _____

15. $\dfrac{3x^4 - 6x^3 + 4x^2 - 3x - 4}{x + 2}$ 16. $\dfrac{3x^4 - 10x^3 - x^2 + 23x - 2}{x - 3}$ 15. _____

                                                                   16. _____

17. $\dfrac{3x^4 - x^2 + 3}{x - 2}$ 18. $\dfrac{x^4 - 4x^2 - 18}{x - 3}$ 17. _____

                                                                   18. _____

Name _____   Score _____

Use the Remainder Theorem to evaluate the polynomial function.

1. $P(x) = 3x^2 - 2x - 5;\ P(3)$

2. $P(x) = x^3 - 5x^2 + 2x + 7;\ P(-2)$

1. _____

2. _____

3. $R(t) = 4t^3 - t^2 + 5t - 11;\ R(-3)$

4. $F(x) = 3x^3 - x + 5;\ F(4)$

3. _____

4. _____

5. $S(t) = 5t^4 - t;\ S(-1)$

6. $R(t) = 4t^3 - 3t + 2t - 5;\ R(2)$

5. _____

6. _____

7. $P(x) = x^4 - 3x^3 + 7x - 5;\ P(5)$

8. $Q(x) = x^5 - 5x^3 - x + 6;\ Q(2)$

7. _____

8. _____

9. $P(t) = t^4 - 4t^3 + 5t^2 - t + 4;\ P(-4)$

10. $Q(y) = 4y^3 + 5y^2 - 7;\ Q(-2)$

9. _____

10. _____

11. $Z(y) = y^4 - y^3 + y^2 - y + 1;\ Z(-1)$

12. $R(x) = x^3 - 4x^2 + 6x - 5;\ R(3)$

11. _____

12. _____

13. $S(x) = x^3 - 3x^2 + 3x - 1;\ S(1)$

14. $R(z) = z^4 - 2z^2 + 4;\ R(-5)$

13. _____

14. _____

15. $F(x) = x^5 + x^4 - x^3 + 44x^2 - x + 3;\ F(-4)$

16. $Y(z) = z^3 + 4z^2 + 4z + 1;\ Y(-3)$

15. _____

16. _____

Name   Score

Factor.

1. $15b^2 + 6b$

2. $3x^3 - 4x^2$

3. $3a^2 - 18b^3$

4. $16x^2 + 18y^2$

5. $x^4 + x^2 - 2x$

6. $3x^3 - 2x^2 + 5x$

7. $2x^5 - x^4 - 4x^2$

8. $x^4 - x^3 + 4x^2$

9. $12x^2 + 9x + 18$

10. $10x^2 - 15x + 20$

11. $20x^2 - 8x + 28$

12. $28x^2 - 21x + 14$

13. $8b^2 + 12b^3 + 16b^4$

14. $30b^4 - 12b^3 + 6b^2$

15. $x^{2n} - x^{3n}$

16. $a^{6n} - a^{5n}$

17. $x^{4n} - x^{2n} + 3x^n$

18. $x^{3n} + x^n$

19. $a^{6n} - a^{3n}$

20. $a^{n+4} - a^4$

21. $15x^2y^2 - 20x^3y + 10x^2y$

22. $40x^3y^3 - 48x^2y^3 + 56x^2y^2$

23. $12x^2y + 24x^2y^2 + 36x^2y^3$

24. $a^{2n+2} - a^{2n+1} + 2a^n$

1.	_____
2.	_____
3.	_____
4.	_____
5.	_____
6.	_____
7.	_____
8.	_____
9.	_____
10.	_____
11.	_____
12.	_____
13.	_____
14.	_____
15.	_____
16.	_____
17.	_____
18.	_____
19.	_____
20.	_____
21.	_____
22.	_____
23.	_____
24.	_____

Name

Score

Factor.

1. $2(x+y)+a(x+y)$

2. $a(x-4)-b(x-4)$

3. $b(a-c)+d(a-c)$

4. $x^2+x+5x+5$

5. $x^2+7x-4x-28$

6. $x^2+6x-2x-12$

7. $x^2+8x-3x-24$

8. $ab+6b-3a-18$

9. $ad+cd-ae-ce$

10. a^2b+4a^2+b+4

11. $10-5x^2+2y-x^2y$

12. $12+3b-4a^2-a^2b$

13. $12+6a^2+2a+a^3$

14. $20+5a^2-4a-a^3$

15. $2ax^2+6ay-6x^2-36y$

16. $4a^2x+6bx-2a^2y-3by$

17. $a^nx^n-3x^n-4a^n+12$

18. $a^nx^n+3a^n+2x^n+6$

1. _____

2. _____

3. _____

4. _____

5. _____

6. _____

7. _____

8. _____

9. _____

10. _____

11. _____

12. _____

13. _____

14. _____

15. _____

16. _____

17. _____

18. _____

Name   Score

Factor.

1. $x^2 - 7x + 10$

2. $x^2 + 11x + 18$

3. $a^2 + 9a + 8$

4. $a^2 - a - 56$

5. $b^2 + 3b - 28$

6. $a^2 + 6a + 5$

7. $a^2 - 8a - 20$

8. $x^2 + 12x + 35$

9. $x^2 + 2x - 63$

10. $a^2 - 8a - 20$

11. $a^2 - 16a + 63$

12. $x^2 + 9x - 10$

13. $a^2 - 14a + 45$

14. $a^2 - 15a - 34$

15. $a^2 + 11a - 26$

16. $a^2 + 7a - 18$

17. $x^2 - 12x + 27$

18. $x^2 + 15x + 56$

19. $x^2 + x - 90$

20. $a^2 - 12a + 27$

21. $x^2 + 14x + 40$

22. $x^2 + 15x + 36$

23. $b^2 - 8b - 48$

24. $x^2 + 7x - 8$

25. $a^2 - 5ab + 4b^2$

26. $a^2 - 9ab + 8b^2$

27. $x^2 + 5xy - 50y^2$

1. _____
2. _____
3. _____
4. _____
5. _____
6. _____
7. _____
8. _____
9. _____
10. _____
11. _____
12. _____
13. _____
14. _____
15. _____
16. _____
17. _____
18. _____
19. _____
20. _____
21. _____
22. _____
23. _____
24. _____
25. _____
26. _____
27. _____

Name   Score

Factor.

1. $2x^2 + 5x + 2$

2. $2x^2 - 5x - 12$

3. $3x^2 + 13x - 10$

4. $3a^2 + 2a - 5$

5. $2a^2 + 11a - 6$

6. $3y^2 - 2y - 16$

7. $6b^2 - 13b + 6$

8. $2x^2 + x - 3$

9. $10x^2 + x - 3$

10. $6x^2 + 29x - 5$

11. $2x^2 + 3x - 5$

12. $5x^2 + 9x - 2$

13. $4x^2 + 13x + 3$

14. $3a^2 + a - 4$

15. $15x^2 + x - 2$

16. $10x^2 - 81x + 8$

17. $12x^2 - x - 6$

18. $4x^2 + 4x - 3$

19. $4x^2 + x - 3$

20. $12y^2 + 7y - 10$

21. $6x^2 - 5yx - 4y^2$

22. $3a^2 + 29ab + 40b^2$

23. $6a^2 + 29ab - 5b^2$

24. $8x^2 + 10xy - 3y^2$

25. $2x^2 + 3x + 5$

1. _____
2. _____
3. _____
4. _____
5. _____
6. _____
7. _____
8. _____
9. _____
10. _____
11. _____
12. _____
13. _____
14. _____
15. _____
16. _____
17. _____
18. _____
19. _____
20. _____
21. _____
22. _____
23. _____
24. _____
25. _____

Name _____   Score _____

Factor.

1. $x^2 - 9$

2. $x^2 - 36$

3. $9x^2 - 1$

4. $9x^2 - 16$

5. $36x^2 - 64$

6. $a^2b^2 - 121$

7. $x^2 + 9$

8. $a^2 + 144$

9. $16 - a^2b^2$

10. $49 - x^2y^2$

11. $36a^2 - 25b^4$

12. $a^{2n} - 9$

13. $b^{2n} - 36$

14. $a^2 + 18a + 81$

15. $9a^2 + 3a - 1$

16. $4x^2 + 8x - 3$

17. $b^2 + 9b + 9$

18. $y^2 - 3y + 24$

19. $x^2 + 4xy + 4y^2$

20. $4x^2y^2 + 29xy + 25$

21. $25a^2 - 20ab + 4b^2$

22. $4a^2 - 20ab + 25b^2$

23. $x^{2n} + 8x^n + 16$

24. $y^{2n} - 20y^n + 100$

1. _____
2. _____
3. _____
4. _____
5. _____
6. _____
7. _____
8. _____
9. _____
10. _____
11. _____
12. _____
13. _____
14. _____
15. _____
16. _____
17. _____
18. _____
19. _____
20. _____
21. _____
22. _____
23. _____
24. _____

Name _____ Score _____

Factor.

1. $x^3 - 8$

2. $x^3 - 64$

3. $27x^3 - 1$

4. $8x^3 + 27$

5. $x^3 + 8y^3$

6. $x^3 + 216y^3$

7. $m^3 - n^3$

8. $125x^3 - y^3$

9. $216a^3 + 1$

10. $1 - 216a^3$

11. $64x^3 - 27y^3$

12. $27x^3 + 8y^3$

13. $x^3y^3 + 8$

14. $27x^3y^3 + 64$

15. $25x^3 - y^3$

16. $8x^3 + 9y^2$

17. $16a^3 - 27$

18. $27x^3 - 125$

19. $125 - x^3$

20. $64 + x^3$

21. $8x^3 - 25$

22. $(a+b)^3 - 8$

23. $a^3 - (a+b)^3$

24. $16x^3 - 25$

25. $x^{3n} - y^{3n}$

26. $y^{3n} + 64$

27. $a^{3n} - 125$

1. _____

2. _____

3. _____

4. _____

5. _____

6. _____

7. _____

8. _____

9. _____

10. _____

11. _____

12. _____

13. _____

14. _____

15. _____

16. _____

17. _____

18. _____

19. _____

20. _____

21. _____

22. _____

23. _____

24. _____

25. _____

26. _____

27. _____

Name

Score

Factor.

1. $x^2y^2 - 8xy - 20$

2. $x^2y^2 - 13xy + 40$

3. $a^2b^2 + 11ab + 28$

4. $a^2b^2 - 15ab + 56$

5. $x^4 - 10x^2 + 16$

6. $y^4 - 11y^2 - 42$

7. $b^4 - 10b^2 + 21$

8. $b^4 + 12b^2 + 27$

9. $a^4 - 7a^2 - 60$

10. $a^4b^4 + 10a^2b^2 - 24$

11. $a^4b^4 + 8a^2b^2 - 33$

12. $a^{2n} - 2a^n - 15$

13. $2x^2y^2 + 3xy - 20$

14. $3x^2y^2 + 2xy - 8$

15. $3a^2b^2 - 17ab + 10$

16. $2a^{2n} + 9a^n + 9$

17. $2a^{2n} - 9a^n - 5$

18. $6a^{2n} + 7a^n + 2$

1. _____

2. _____

3. _____

4. _____

5. _____

6. _____

7. _____

8. _____

9. _____

10. _____

11. _____

12. _____

13. _____

14. _____

15. _____

16. _____

17. _____

18. _____

Name _____

Score _____

Factor.

1. $4x^2 - 8x + 4$

2. $5x^2 - 20$

3. $6x^2 - 6$

4. $16x^2 - 36$

5. $x^4 - 81$

6. $81ax^4 - 1$

7. $12x^4 - 75x^2$

8. $3a - 3a^4$

9. $a^6 b^3 + 27a^3$

10. $3x^4 + 9x^3 - 120x^2$

11. $x^4 - 3x^3 - 28x^2$

12. $a^4 + 5a^2 - 36$

13. $a^4 - 8a^2 - 9$

14. $3a^4 - 24a^3 + 48a^2$

15. $a^4 - 5a^2 + 4$

16. $x^4 - 16$

17. $10x^2 - 5x - 15$

18. $6y^2 + 30y + 36$

19. $15x^3 y - 18x^2 y^2 - 24xy^3$

20. $16x^4 + 4x^3 y - 30x^2 y^2$

21. $a^{2n+3} - 8a^{n+2} + 16a^2$

22. $x^{2n+1} + 4x^{n+1} + 4x$

23. $2x^{n-2} - 7x^{n+1} + 6x^n$

24. $2a^{n+2} + 5a^{n+1} - 3a^n$

1. _____
2. _____
3. _____
4. _____
5. _____
6. _____
7. _____
8. _____
9. _____
10. _____
11. _____
12. _____
13. _____
14. _____
15. _____
16. _____
17. _____
18. _____
19. _____
20. _____
21. _____
22. _____
23. _____
24. _____

Name   Score

Solve by factoring.

1. $(x-4)(x+6) = 0$ 2. $(3x+4)(x-7) = 0$ 1. _____

                                                                2. _____

3. $4x(5x-2)(3x-8) = 0$ 4. $x(x-6)(3x+1) = 0$ 3. _____

                                                                4. _____

5. $7x^2 - 21x = 0$ 6. $5x^2 = 15x$ 5. _____

                                                                6. _____

7. $25x^2 = 4$ 8. $z^2 - 9 = 0$ 7. _____

                                                                8. _____

9. $x^2 - 12 = x$ 10. $x^2 = 2x + 15$ 9. _____

                                                                10. _____

11. $x^3 + 3x^2 + 2x = 0$ 12. $x^4 - 8x^2 + 16 = 0$ 11. _____

                                                                12. _____

13. $x^3 - x^2 - 9x + 9 = 0$ 14. $3x^3 - 7x^2 - 12x + 28 = 0$ 13. _____

                                                                14. _____

15. $4x^3 - 3x^2 - 4x + 3 = 0$ 16. $4x^2 + 8x - 26 = (x+3)(x-2)$ 15. _____

                                                                16. _____

Name _____ Score _____

Solve.

1. The sum of a number and its square is 42. Find the number.

2. The sum of a number and its square is 90. Find the number.

    1. _____

    2. _____

3. The length of a rectangle is 9 cm more than four times the width. The area of the rectangle is 198 cm². Find the length and the width.

4. The height of a triangle is 6 cm more than twice the length of the base. The area of the triangle is 88 cm². Find the base and the height of the triangle.

    3. _____

    4. _____

5. One leg of a right triangle is 3 ft more than three times the other leg. The hypotenuse is 1 ft more than the longer leg. Find the length of the hypotenuse of the right triangle.

6. The length of a rectangle is 6 ft more than five times the width. The area of the rectangle is 287 ft². Find the length and width.

    5. _____

    6. _____

7. An object is thrown downward with an initial speed of 16 ft/s from the top of a tower 192 ft high. How many seconds later will the object hit the ground? Use the equation $d = vt + 16t^2$, where d is the distance in feet, v is the initial speed, and t is the time in seconds.

8. A penny is thrown into a wishing well with an initial speed of 8 ft/s. The well is 48 ft deep. How many seconds later will the penny hit the bottom of the well? Use the equation $d = vt + 16t^2$, where d is the distance in feet, v is the initial speed, and t is the time in seconds.

    7. _____

    8. _____

Name   Score

Evaluate.

1.   Given $f(x) = \dfrac{-7x}{x^2 - 3x + 6}$, find $f(2)$.

2.   Given $g(x) = \dfrac{x+8}{9x^2 - 6x + 3}$, find $g(-2)$.

3.   Given $f(x) = \dfrac{x-8}{3x^2 - 2x - 11}$, find $f(-3)$.

4.   Given $g(x) = \dfrac{-6}{x^2 + 5x - 9}$, find $g(4)$.

5.   Given $f(x) = \dfrac{x+14}{2x^2 - 7x + 2}$, find $f(-5)$.

6.   Given $g(x) = \dfrac{-12}{x^2 - 5x - 4}$, find $g(3)$.

7.   Given $f(x) = \dfrac{5}{x^2 - 9x + 1}$, find $f(4)$.

8.   Given $g(x) = \dfrac{x+3}{3x^2 + 4x - 5}$, find $g(1)$.

Find the domain.

9. $f(x) = \dfrac{x-7}{2x+6}$

10. $g(x) = \dfrac{x+11}{x^2 - x - 12}$

11. $f(x) = \dfrac{6x+5}{x^2 + 2x + 1}$

12. $g(x) = \dfrac{4x-3}{5x - 35}$

13. $f(x) = \dfrac{6}{x^2 - 9}$

14. $g(x) = \dfrac{5x+3}{x^2 - 3x - 10}$

15. $f(x) = \dfrac{-41}{6x+24}$

16. $g(x) = \dfrac{14x-3}{x^2 - x - 6}$

1.   _____

2.   _____

3.   _____

4.   _____

5.   _____

6.   _____

7.   _____

8.   _____

9.   _____

10. _____

11. _____

12. _____

13. _____

14. _____

15. _____

16. _____

Name _____   Score _____

Simplify.

1. $\dfrac{3+9x}{3}$

2. $\dfrac{12x^2-4x}{4x}$

3. $\dfrac{10x^2(x+2)}{2x(x+2)}$

4. $\dfrac{20x^4(x+7)}{15x^3(x+7)}$

5. $\dfrac{3x-9}{3x-x^2}$

6. $\dfrac{12x^3-6x^2+18x}{6x}$

7. $\dfrac{4x^{3n}-12x^{2n}}{16x^{2n}}$

8. $\dfrac{a^2-8a+15}{a^2+2a-35}$

9. $\dfrac{x^2-5x}{x^2+5x}$

10. $\dfrac{x^{3n}+x^{2n}y^n}{x^{3n}-x^{2n}y^n}$

11. $\dfrac{9a^n}{3a^{2n}-9a^n}$

12. $\dfrac{12-x-x^2}{2x^2-7x+3}$

13. $\dfrac{3x^2-13x-10}{10-17x+3x^2}$

14. $\dfrac{x^2-3xy-4y^2}{x^2+4xy+3y^2}$

15. $\dfrac{x^2+7x-18}{x^2-5x+6}$

16. $\dfrac{27x^3-y^3}{9x^2-y^2}$

17. $\dfrac{6x^2y^2+7xy^2+2}{6x^2y^2+19xy+10}$

18. $\dfrac{x^2-9}{a(x+3)-b(x+3)}$

19. $\dfrac{x^4-7x^2-8}{x^4+9x^2+8}$

20. $\dfrac{x^4-16}{x^4-x^2-12}$

21. $\dfrac{x^2y^2+8xy-36}{x^2y^2+2xy-63}$

22. $\dfrac{a^{2n}-3a^n-28}{a^{2n}-a^n-42}$

23. $\dfrac{a^{2n}-a^n-6}{a^{2n}+2a^n-15}$

1. _____

2. _____

3. _____

4. _____

5. _____

6. _____

7. _____

8. _____

9. _____

10. _____

11. _____

12. _____

13. _____

14. _____

15. _____

16. _____

17. _____

18. _____

19. _____

20. _____

21. _____

22. _____

23. _____

Name   Score

Multiply.

1. $\dfrac{30a^3b^5}{25x^3y^3} \cdot \dfrac{5x^2y^3}{6a^2b^3}$

2. $\dfrac{18x^4y^5}{16a^2b^3} \cdot \dfrac{32a^3b^2}{36x^3y^5}$

3. $\dfrac{x^3y^2}{x^2-6x-7} \cdot \dfrac{2x^2-15x+7}{x^3y^4}$

4. $\dfrac{2x^2-x-10}{x^3y^4} \cdot \dfrac{x^5y^2}{2x^2-3x-5}$

5. $\dfrac{x^2+2x-15}{x^2+3x-10} \cdot \dfrac{x^2-x-2}{x^2-1}$

6. $\dfrac{x^2-x-6}{15-8x+x^2} \cdot \dfrac{x^2-x-20}{x^2+4x+4}$

7. $\dfrac{x^2+4x-12}{x^2-7x+6} \cdot \dfrac{x^2-3x+2}{x^2-4x+4}$

8. $\dfrac{9x^2+3x-2}{6x^2+13x-5} \cdot \dfrac{2x^2-13x+20}{8+10x-3x^2}$

9. $\dfrac{x^{n+1}+3x^n}{6x^2-9x} \cdot \dfrac{8x-12}{x^{n+1}-x^n}$

10. $\dfrac{x^{2n}+4x^n}{x^{n+1}+4x} \cdot \dfrac{x^2-5x}{x^{n+1}-5x^n}$

11. $\dfrac{2x^2-x-28}{2x^2-11x+15} \cdot \dfrac{3x^2-14x+15}{2x^2-x-28}$

12. $\dfrac{2x^2+13x+15}{2x^2+3x-35} \cdot \dfrac{6x^2-29x-28}{6x^2+x-12}$

13. $\dfrac{2x^2-5x-3}{4x^2-1} \cdot \dfrac{6x^2-3x}{x^2-6x+9}$

14. $\dfrac{6x^2-13x-28}{6x^2-x-12} \cdot \dfrac{2x^2-13x+15}{2x^2+3x-35}$

15. $\dfrac{x^{2n}-2x^n-8}{x^{2n}-5x^n+6} \cdot \dfrac{x^{2n}+2x^n-15}{x^{2n}+x^n-20}$

16. $\dfrac{x^{4n}-16}{x^{2n}+7x^n+10} \cdot \dfrac{x^{2n}+3x^n-10}{x^{2n}+4}$

17. $\dfrac{x^3-8}{2x^2-7x+6} \cdot \dfrac{2x^2+x-6}{x^2+2x+4}$

18. $\dfrac{x^2-6x+5}{3x^2-5x+2} \cdot \dfrac{3x^2-8x+4}{x^2-25}$

1. _____

2. _____

3. _____

4. _____

5. _____

6. _____

7. _____

8. _____

9. _____

10. _____

11. _____

12. _____

13. _____

14. _____

15. _____

16. _____

17. _____

18. _____

Name _____   Score _____

Divide.

1. $$\frac{6x^2y^4}{15a^4b^3} \div \frac{16x^4y^3}{5a^3b^3}$$

2. $$\frac{9a^4b^7}{13x^3y^2} \div \frac{27a^5b^4}{39x^2y}$$

3. $$\frac{4x-12}{10x^2+15x} \div \frac{6x^3-18x^2}{14x^4+21x^3}$$

4. $$\frac{8x^2-8y^2}{4x^2y^2} \div \frac{6x^2+6xy}{2x^2-2xy^2}$$

5. $$\frac{x^4y^3}{x^2+4x-5} \div \frac{x^2y^2}{2x^2+11x+5}$$

6. $$\frac{2x^2-13x+6}{x^3y^3} \div \frac{2x^2-9x-18}{x^4y^4}$$

7. $$\frac{x^2-4x+3}{x^2+5x-6} \div \frac{x^2-2x-3}{x^2-x-2}$$

8. $$\frac{x^2+3x-18}{12-x-x^2} \div \frac{x^2+2x-24}{x^2+10x+24}$$

9. $$\frac{x^{n+1}-2x^n}{3x^2-6x} \div \frac{x^{n+1}+2x^n}{6x^2+12x}$$

10. $$\frac{2x^2+x-45}{3x^2+4x} \div \frac{x^2+2x-15}{12+5x-3x^2}$$

11. $$\frac{3x^2-11x+10}{3x^2+x-10} \div \frac{2x^2-15x+28}{2x^2-3x-14}$$

12. $$\frac{2x^2+x-15}{2x^2-9x+10} \div \frac{3x^2+5x-12}{6x^2+10x-24}$$

13. $$\frac{4x^2-25}{2x^2-3x-20} \div \frac{6x^2+3x-45}{6x^2-6x-72}$$

14. $$\frac{x^2+3x+2}{6-7x-3x^2} \div \frac{x^2-9x-7}{14-23x+3x^2}$$

15. $$\frac{x^{2n}-2x^n-15}{x^{2n}+2x^n-3} \div \frac{x^{2n}-3x^n-4}{x^{2n}-1}$$

16. $$\frac{x^{2n}+5x^n+6}{x^{2n}-x^n-6} \div \frac{x^{2n}+x^n-6}{x^{2n}+6x^n+9}$$

17. $$\frac{x^3+8}{2x^2+x-6} \div \frac{x^2-2x+4}{2x^2+7x-15}$$

18. $$\frac{x^4-10x^2+9}{3x^2-x-2} \div \frac{x^2-9}{3x^2-10x-8}$$

1. _____

2. _____

3. _____

4. _____

5. _____

6. _____

7. _____

8. _____

9. _____

10. _____

11. _____

12. _____

13. _____

14. _____

15. _____

16. _____

17. _____

18. _____

Name _____   Score _____

Write each fraction in terms of the LCM of the denominators.

1. $\dfrac{x-3}{2x(x-3)}$, $\dfrac{2}{6x^2}$

2. $\dfrac{2x-3}{4x(2x-1)}$, $\dfrac{3}{4x^3}$

3. $\dfrac{3x-2}{4x^2-12x}$, $-4x$

4. $\dfrac{5x-4}{2x(x-4)}$, $3x$

5. $\dfrac{6}{5x+3}$, $\dfrac{-8}{5x-3}$

6. $\dfrac{5x}{x^2-4}$, $\dfrac{x+3}{x-2}$

7. $\dfrac{2x}{9-x^2}$, $\dfrac{4x}{9-3x}$

8. $\dfrac{5}{2x^2-8y^2}$, $\dfrac{3}{3x-6y}$

9. $\dfrac{3x}{x^2-25}$, $\dfrac{x-2}{5x-25}$

10. $\dfrac{2x}{x^2-1}$, $\dfrac{3x}{x^2+2x+1}$

11. $\dfrac{x^2+1}{x^3-8}$, $\dfrac{3}{x^2+2x+4}$

12. $\dfrac{x-4}{64-x^3}$, $\dfrac{3}{16+4x+x^2}$

13. $\dfrac{7x}{x^2-x-6}$, $\dfrac{-4x}{x^2+4x+4}$

14. $\dfrac{5x}{4x^2-4x-15}$, $\dfrac{-3x}{4x^2+16x+15}$

15. $\dfrac{-3x}{2x^2-3x-2}$, $\dfrac{2x}{2x^2-5x-3}$

16. $\dfrac{5x}{4x^2-4x-15}$, $\dfrac{-3x}{4x^2+16x+15}$

17. $\dfrac{4}{2x^2+5x-3}$, $\dfrac{3x}{1-2x}$, $\dfrac{2x+1}{x+3}$

18. $\dfrac{6}{3x^2-14x+8}$, $\dfrac{x}{4-x}$, $\dfrac{x-1}{3x-2}$

19. $\dfrac{2x}{x-5}$, $\dfrac{4}{x+2}$, $\dfrac{x+3}{10+3x-x^2}$

20. $\dfrac{3x}{x-2}$, $\dfrac{-1}{x+3}$, $\dfrac{x-2}{6-x-x^2}$

21. $\dfrac{3}{x^{2n}-4}$, $\dfrac{6}{x^{2n}+4x^n+4}$

22. $\dfrac{x-3}{x^{2n}+x^n-2}$, $\dfrac{5x}{x^n-1}$

1. _____
2. _____
3. _____
4. _____
5. _____
6. _____
7. _____
8. _____
9. _____
10. _____
11. _____
12. _____
13. _____
14. _____
15. _____
16. _____
17. _____
18. _____
19. _____
20. _____
21. _____
22. _____

Name

Score

Simplify.

1. $\dfrac{2}{3xy} - \dfrac{5}{3xy} + \dfrac{3}{3xy}$

2. $-\dfrac{4}{5x^2} + \dfrac{9}{5x^2} - \dfrac{4}{5x^2}$

3. $\dfrac{x}{x^2 - 4x - 5} - \dfrac{5}{x^4 - 4x - 5}$

4. $\dfrac{3x}{3x^2 + 13x - 10} - \dfrac{2}{3x^2 + 13x - 10}$

5. $\dfrac{x}{x^2 - y^2} - \dfrac{y}{x^2 - y^2}$

6. $\dfrac{4x}{4x^2 + 3x - 10} - \dfrac{5}{4x^2 + 3x - 10}$

7. $\dfrac{3}{5xy^2} - \dfrac{4}{15y} - \dfrac{7}{3xy}$

8. $\dfrac{3}{4a^2b} - \dfrac{5}{8ab^2} - \dfrac{1}{12a^2b^2}$

9. $\dfrac{3x - 2}{15x} - \dfrac{2x - 5}{10x}$

10. $\dfrac{5a}{a - 2} - 4 + \dfrac{2}{a}$

11. $\dfrac{3x - 4}{x + 5} - \dfrac{x^2 - 8x - 20}{x^2 + 9x + 20}$

12. $\dfrac{2}{x + 5} - \dfrac{4x}{x^2 + 10x + 25}$

13. $\dfrac{x}{x + 5} - \dfrac{5 - x}{x^2 - 25}$

14. $\dfrac{1}{x + 4} - \dfrac{3x}{x^2 + 8x + 16}$

15. $\dfrac{5x}{x - 4} - \dfrac{4x}{x - 5}$

16. $\dfrac{6a}{a + 4} - \dfrac{11a}{a - 1}$

17. $\dfrac{x - 2}{x + 4} - \dfrac{x^2 - 3x - 18}{x^2 + 7x + 12}$

18. $\dfrac{3x^n - 9}{x^{2n} + x^n - 12} - \dfrac{x^n}{x^n - 3}$

19. $\dfrac{3x - 2}{4x^2 - 9} - \dfrac{4}{2x - 3}$

20. $\dfrac{x^2 + 3}{5x^2 - 20} - \dfrac{8}{x + 2}$

1. _____

2. _____

3. _____

4. _____

5. _____

6. _____

7. _____

8. _____

9. _____

10. _____

11. _____

12. _____

13. _____

14. _____

15. _____

16. _____

17. _____

18. _____

19. _____

20. _____

Name   Score

Simplify.

1. $\dfrac{3+\dfrac{1}{x}}{9-\dfrac{1}{x^2}}$

2. $\dfrac{\dfrac{16}{x^2}-1}{1+\dfrac{4}{x}}$

3. $\dfrac{a-5}{\dfrac{25}{a}-a}$

4. $\dfrac{\dfrac{36}{a}-a}{6+a}$

5. $\dfrac{\dfrac{3}{a^2}+\dfrac{3}{a}}{\dfrac{3}{a^2}+\dfrac{3}{a}}$

6. $\dfrac{\dfrac{1}{x}+\dfrac{1}{4}}{\dfrac{16}{x^2}-1}$

7. $\dfrac{1+\dfrac{1}{x}-\dfrac{12}{x^2}}{1+\dfrac{3}{x}-\dfrac{18}{x^2}}$

8. $\dfrac{\dfrac{12}{x^2}-\dfrac{5}{x}-2}{\dfrac{8}{x^2}-\dfrac{10}{x}+3}$

9. $\dfrac{6+\dfrac{13}{x}-\dfrac{5}{x^2}}{2-\dfrac{11}{x}+\dfrac{15}{x^2}}$

10. $\dfrac{2+\dfrac{18}{x}+\dfrac{16}{x^2}}{5+\dfrac{2}{x}-\dfrac{3}{x^2}}$

11. $\dfrac{1-\dfrac{5}{2x-3}}{x-\dfrac{24}{2x-3}}$

12. $\dfrac{1+\dfrac{4}{3x+5}}{x-\dfrac{12}{3x+5}}$

13. $\dfrac{x+2-\dfrac{18}{x-1}}{x+12+\dfrac{42}{x-1}}$

14. $\dfrac{x+2-\dfrac{15}{x+4}}{x+5-\dfrac{6}{x+4}}$

15. $\dfrac{\dfrac{2}{a}-\dfrac{4}{a-3}}{\dfrac{1}{a}+\dfrac{3}{a-3}}$

16. $\dfrac{\dfrac{3}{a}-\dfrac{6}{a+4}}{\dfrac{4}{a}+\dfrac{4}{a+4}}$

17. $\dfrac{\dfrac{x-2}{x+2}-\dfrac{x+2}{x-2}}{\dfrac{x-2}{x+2}+\dfrac{x+2}{x-2}}$

18. $\dfrac{\dfrac{x}{x+4}-\dfrac{x}{x-4}}{\dfrac{x}{x+4}+\dfrac{x}{x-4}}$

1. _____

2. _____

3. _____

4. _____

5. _____

6. _____

7. _____

8. _____

9. _____

10. _____

11. _____

12. _____

13. _____

14. _____

15. _____

16. _____

17. _____

18. _____

Name   Score

Solve.

1. $\dfrac{x}{18} = \dfrac{2}{3}$

2. $\dfrac{12}{16} = \dfrac{x}{48}$

1. _____

2. _____

3. $\dfrac{4}{x} = \dfrac{8}{20}$

4. $\dfrac{5}{x} = \dfrac{8}{40}$

3. _____

4. _____

5. $\dfrac{x+3}{18} = \dfrac{1}{2}$

6. $\dfrac{x+5}{12} = \dfrac{3}{4}$

5. _____

6. _____

7. $\dfrac{6-x}{15} = \dfrac{5}{3}$

8. $\dfrac{10-x}{8} = \dfrac{7}{2}$

7. _____

8. _____

9. $\dfrac{5}{x+3} = \dfrac{1}{2}$

10. $\dfrac{x}{2} = \dfrac{x-3}{4}$

9. _____

10. _____

11. $\dfrac{-2}{x-3} = \dfrac{5}{x}$

12. $\dfrac{5}{2} = \dfrac{20}{x+2}$

11. _____

12. _____

13. $\dfrac{x}{5} = \dfrac{x-8}{15}$

14. $\dfrac{12}{4-x} = \dfrac{4}{x}$

13. _____

14. _____

15. $\dfrac{6}{x+1} = \dfrac{4}{x-1}$

16. $\dfrac{10}{x-2} = \dfrac{15}{x-1}$

15. _____

16. _____

17. $\dfrac{x}{5} = \dfrac{x+2}{10}$

18. $\dfrac{x}{2} = \dfrac{x-1}{6}$

17. _____

18. _____

Name   Score

Solve.

1. The real estate tax for a house that cost $75,000 is $1600. At this rate, what is the value of a house for which the real estate tax is $1920?

2. The license fee for a car that cost $8000 is $96. At the same rate, what is the license fee for a car that cost $9600?

1. _____

2. _____

3. A pre-election survey showed that 4 out of every 9 voters would vote in a special election. At this rate, how many people would be expected to vote in a city of 513,000?

4. A quality control inspector found 7 defective DVDs in a shipment of 4000 DVDs. At this rate, how many DVDs would be expected to be defective in a shipment of 12,000 DVDs?

3. _____

4. _____

5. The scale on an architectural drawing is $\frac{1}{8}$ in. representing 1 ft. Find the dimensions of a room that measures $9\frac{1}{2}$ in. by 12 in. on the drawing.

6. Two pounds of pecans cost $7.50. At this rate, how much would 25 lb of pecans cost? Round to the nearest cent.

5. _____

6. _____

7. One and one-fourth ounces of a medication are required for a 120-pound adult. At the same rate, how many additional ounces of medication are required for a 180-pound adult?

8. Ten ounces of an insecticide is mixed with 32 gal of water to make a spray for spraying a cotton field. How much additional insecticide is required to be mixed with 80 gal of water?

7. _____

8. _____

9. A stock investment of 400 shares pays a dividend of $640. At this rate, how many additional shares are required to earn a dividend of $1000?

10. An investment of $5000 earns $600 each year. At this rate, how much additional money must be invested to earn $1440 each year?

9. _____

10. _____

Name _____   Score _____

Solve.

1. $\dfrac{x}{3} + \dfrac{3}{4} = \dfrac{x}{12}$

2. $\dfrac{x}{5} - \dfrac{3}{8} = \dfrac{x}{20}$

3. $\dfrac{6}{5x-3} = 3$

4. $-2 = \dfrac{10}{3x+4}$

5. $3 - \dfrac{4}{x} = 5$

6. $5 - \dfrac{8}{x} = 7$

7. $\dfrac{1}{x-2} = \dfrac{3}{x}$

8. $\dfrac{8}{x} = \dfrac{4}{x+1}$

9. $\dfrac{8}{x-3} = \dfrac{5}{x}$

10. $\dfrac{4}{x} = \dfrac{2}{x-6}$

11. $6 + \dfrac{9}{a-4} = \dfrac{3a}{a-4}$

12. $\dfrac{-6}{a-6} = 4 - \dfrac{a}{a-6}$

13. $-\dfrac{4}{x+8} + 2 = \dfrac{6}{x+8}$

14. $7 - \dfrac{3}{2x-3} = \dfrac{4}{2x-3}$

15. $\dfrac{3}{x^2-25} + \dfrac{1}{x+5} = \dfrac{8}{x-5}$

16. $\dfrac{6}{x-4} - \dfrac{3}{x+4} = \dfrac{9}{x^2-16}$

17. $\dfrac{11}{x^2-5x+6} = \dfrac{3}{x-2} + \dfrac{4}{x-3}$

18. $\dfrac{4}{x^2+8x+12} = \dfrac{7}{x+2} - \dfrac{3}{x+6}$

1. _____

2. _____

3. _____

4. _____

5. _____

6. _____

7. _____

8. _____

9. _____

10. _____

11. _____

12. _____

13. _____

14. _____

15. _____

16. _____

17. _____

18. _____

Name   Score

Solve.

1. One printer can print the paychecks
    for the employees of a company in
    60 min. A second printer can print the
    checks in 90 min. How long would it
    take to print the checks with both printers
    operating?

2. One solar heating panel can raise the
    temperature of water 1° in 30 min. A
    second solar heating panel can raise
    the temperature 1° in 45 min. How
    long would it take to raise the temperature
    of the water 1° with both solar panels
    operating?

3. A new printer can print checks three
    times faster than an old printer. The
    older printer can print the checks in
    48 min. How long would it take to
    print the checks with both printers
    operating?

4. An experienced electrician can wire
    a room twice as fast as an apprentice
    electrician. Working together, the
    electricians can wire a room in 4 h.
    How long would it take the apprentice
    working along to wire a room?

5. One member of a gardening team
    can mow and clean up a lawn in 12 h.
    With both members of the team
    working, the job can be done in 8 h.
    How long would it take the second
    member of the team, working alone
    to do the job?

6. A welder requires 18 h to do a job.
    After the welder and an apprentice
    work on a job for 6 h, the welder
    moves to another job. The apprentice
    finishes the job in 14 h. How long
    would it take the apprentice, working
    alone, to do the job?

7. Three computers can print out a task
    in 12 min, 18 min, and 36 min,
    respectively. How long would it take
    to complete the task with all three
    computers working?

8. Three machines are filling soda
    bottles. The machines can fill the
    daily quota of soda bottles in 10 h,
    15 h, and 30 h, respectively. How
    long would it take to fill the daily
    quota of soda bottles with all three
    machines working?

9. The inlet pipe can fill a water tank
    in 48 min. The outlet pipe can empty
    the tank in 16 min. How long does it
    take to empty a full tank when both
    pipes are open?

10. An oil tank has two inlet pipes and
    one outlet pipe. One inlet pipe can
    fill the tank in 6 h, and the other inlet
    pipe can fill the tank in 12 h. The outlet
    pipe can empty the tank in 24 h. How
    long would it take to fill the tank with
    all three pipes open.

1. _____

2. _____

3. _____

4. _____

5. _____

6. _____

7. _____

8. _____

9. _____

10. _____

Name _____ Score _____

Solve.

1. An express bus travels 455 mi in the same amount of time as a car travels 406 mi. The rate of the car is 7 mph less than the rate of the bus. Find the rate of the car.

1. _____

2. A commercial jet travels 2700 mi in the same amount of time as a corporate jet travels 2175 miles. The rate of the commercial jet is 105 mph faster than the rate of the corporate jet. Find the rate of each jet.

2. _____

3. A motorcycle travels 232 mi in the same amount of time as a car travels 200 mi. The rate the motorcycle is 8 mph faster than the rate of the car. Find the rate of the car.

3. _____

4. A sales executive traveled 75 mi by car and then an additional 1400 mi by plane. The rate of the plane was eight times faster than the rate of the car. The total time for the trip was 5 h. Find the rate of the plane.

4. _____

5. An express train and a car leave a town at 2 P.M. and head for a town 300 mi away. The rate of the express is one and one-half times the rate of the car. The train arrives 2 h ahead of the car. Find the rate of the car.

5. _____

6. A corporate jet and a car start from a town at 7 A.M. and head for a town 660 mi away. The rate of the plane is six times the rate of the car. The plane arrives 10 h ahead of the car. Find the rate of the car.

6. _____

7. A plane can fly at a rate of 212 mph in calm air. Traveling with the wind, the plane flew 720 in the same amount of time as it flew 552 mi against the wind. Find the rate of the wind.

7. _____

8. A jet can fly at a rate of 500 mph in calm air. Traveling with the wind, the plane flew 2240 mi in the same amount of time as it flew 1760 mi against the wind. Find the rate of the wind.

8. _____

9. A passenger train travels 165 mi in the same amount of time as a freight train travel 120 mi. The rate of the passenger train is 15 mph faster than the rate of the freight train. Find the rate of each train.

9. _____

10. A plane can fly at a rate of 240 mph in calm air. Traveling with the wind, the plane flew 900 mi in the same amount of time as it flew 540 mi against the wind. Find the rate of the wind.

10. _____

Name _____   Score _____

Solve.

1. The distance (d) a spring will stretch varies directly as the force (f) applied to the spring. If a force of 12 lb is required to stretch a spring 8 in., what force is required to stretch the spring 12 in.?

2. The pressure (p) on a diver in the water varies directly as the depth (d). If the pressure is 8 lb/in.^{2} when the depth is 16 ft, what is the pressure when the depth is 20 ft?

1. _____

2. _____

3. The stopping distance (s) of a car varies directly as the square of its speed (v). If a car traveling 40 mph requires 150 ft to stop, find the stopping distance for a car traveling 50 mph.

4. The distance (s) a ball will roll down an inclined plane is directly proportional to the square of the time (t). If a ball rolls 8 ft in 1 s, how far will it roll in 4 s?

3. _____

4. _____

5. The speed (v) of a gear varies inversely as the number of teeth (t). If a gear that has 30 teeth makes 25 revolutions per minute, how many revolutions per minute will a gear that has 50 teeth make?

6. For a constant temperature, the pressure (P) of a gas varies inversely as the volume (V). If the pressure is 70 lb/in^2 when the volume is 210 ft^3, find the pressure when the volume is 140 ft^3.

5. _____

6. _____

7. The intensity (I) of a light source is inversely proportional to the square of the distance (d) from the source. If the intensity is 48 lumens at a distance of 12 ft, what is the intensity at a distance of 6 ft?

8. The length (L) of a rectangle of fixed area varies inversely as the width (W). If the length of a rectangle is 10 ft when the width is 8 ft, find the length of the rectangle when the width if 5 ft.

7. _____

8. _____

9. The frequency of vibration (f) in an open pipe organ varies inversely as the length (L) of the pipe. If the air in a pipe 3 m long vibrates 90 times per minute, find the frequency in a pipe that is 2 m long.

10. The current (I) in a wire varies directly as the voltage (v) and inversely as the resistance (r). If the current is 32 amps when the voltage is 192 volts and the resistance is 6 ohms, find the current when the voltage is 210 volts and the resistance is 12 ohms.

9. _____

10. _____

Name _____   Score _____

Simplify.

1. $25^{1/2}$

2. $27^{1/3}$

3. $16^{3/2}$

4. $16^{-1/2}$

5. $(-49)^{3/2}$

6. $(-27)^{1/3}$

7. $\left(\dfrac{16}{49}\right)^{-1/2}$

8. $x^{-3/5} \cdot x^{1/5}$

9. $a^{1/3} \cdot a^{4/3}$

10. $x \cdot x^{-1/4}$

11. $\dfrac{a^{3n}}{a^{-n}}$

12. $\left(a^{-1/4}\right)^{-4}$

13. $x^{2/5} x^{-3/5}$

14. $\left(a^4 b^8\right)^{3/4}$

15. $\left(x^{-6} y^9\right)^{-1/3}$

16. $\left(\dfrac{x^{1/3}}{y^{-2}}\right)^6$

17. $x^{2/3}\left(x^{2/3} - x^{-1/3}\right)$

18. $\dfrac{a^{3/2}}{a^{1/2}}$

19. $\dfrac{x^{3n}}{x^n}$

20. $a^{n/2} a^{-n/4}$

21. $\dfrac{x^{-2/5}}{x^{1/5}}$

22. $a^{2n} a^{-4n}$

23. $\dfrac{x^{3/4}}{x^{-1/4}}$

24. $\left(x^{-3}\right)^{1/3}$

25. $\left(a^6\right)^{-2/3}$

26. $\left(x^{-2/3}\right)^9$

27. $\left(x^{-3/4}\right)^{12}$

1. _____
2. _____
3. _____
4. _____
5. _____
6. _____
7. _____
8. _____
9. _____
10. _____
11. _____
12. _____
13. _____
14. _____
15. _____
16. _____
17. _____
18. _____
19. _____
20. _____
21. _____
22. _____
23. _____
24. _____
25. _____
26. _____
27. _____

Name   Score

Rewrite the exponential expression as a radical expression.

1. $2^{1/3}$ 2. $7^{1/2}$ 3. $b^{3/5}$

1. _____

2. _____

3. _____

4. $(2x)^{7/4}$ 5. $-9a^{2/3}$ 6. $\left(ab^2\right)^{1/3}$

4. _____

5. _____

6. _____

7. $\left(x^2 y^2\right)^{3/5}$ 8. $\left(a^2 b^2\right)^{2/3}$ 9. $(3a-2b)^{1/2}$

7. _____

8. _____

9. _____

10. $(3x-4)^{1/3}$ 11. $(2x-1)^{2/3}$ 12. $x^{-3/4}$

10. _____

11. _____

12. _____

Rewrite the exponential expression as a radical expression.

13. $\sqrt{15}$ 14. $\sqrt[3]{y}$ 15. $\sqrt[4]{5y^5}$

13. _____

14. _____

15. _____

16. $-\sqrt{2x^3}$ 17. $-\sqrt[3]{3x^5}$ 18. $-\sqrt[4]{2x^7}$

16. _____

17. _____

18. _____

19. $-\sqrt[3]{4x^4}$ 20. $2x\sqrt[3]{y^2}$ 21. $3x\sqrt[4]{y}$

19. _____

20. _____

21. _____

22. $\sqrt{a^2-8}$ 23. $\sqrt{2-x^2}$ 24. $\sqrt{x^2+y^2}$

22. _____

23. _____

24. _____

Name _____ Score _____

Simplify.

1. $\sqrt{x^{12}}$

2. $\sqrt{x^{10}}$

3. $-\sqrt{x^{14}}$

4. $-\sqrt{a^4}$

5. $-\sqrt[3]{x^{12}y^6}$

6. $-\sqrt[3]{a^6b^{15}}$

7. $-\sqrt[3]{a^3b^9}$

8. $\sqrt[5]{x^{10}y^{25}}$

9. $\sqrt[4]{x^{12}y^{20}}$

10. $\sqrt[3]{-a^9b^{12}}$

11. $\sqrt{144x^{12}y^{16}}$

12. $\sqrt{9a^4b^{10}}$

13. $\sqrt{81x^6y^4}$

14. $\sqrt[3]{64x^6}$

15. $\sqrt[3]{125x^{12}y^3}$

16. $\sqrt[3]{-x^{12}y^{21}}$

17. $\sqrt[3]{-8a^{18}b^9}$

18. $-\sqrt[4]{a^8b^{12}}$

19. $\sqrt[5]{x^{20}y^{30}}$

20. $\sqrt[5]{x^{15}y^{25}}$

21. $\sqrt[4]{16a^{16}b^{20}}$

22. $\sqrt[4]{625a^8b^{12}}$

23. $\sqrt[5]{-a^5b^{15}}$

24. $\sqrt[5]{243x^{10}y^{15}}$

1. _____

2. _____

3. _____

4. _____

5. _____

6. _____

7. _____

8. _____

9. _____

10. _____

11. _____

12. _____

13. _____

14. _____

15. _____

16. _____

17. _____

18. _____

19. _____

20. _____

21. _____

22. _____

23. _____

24. _____

Name _____   Score _____

Simplify.

1. $\sqrt{x^4 y^6 z^7}$

2. $\sqrt{98a^4 b^6}$

3. $\sqrt{50xy^6 z^9}$

4. $\sqrt{75x^3 y^4 z^6}$

5. $\sqrt{-16x^5}$

6. $\sqrt{-x^3 y^6}$

7. $\sqrt{x^5 y^7 z^9}$

8. $\sqrt{x^6 y^8 z^4}$

9. $\sqrt{x^6 z^{12}}$

10. $\sqrt{192xy^7 z^{10}}$

11. $\sqrt{128x^3 y^5 z^6}$

12. $\sqrt[3]{8x^6 y^9 z^{10}}$

13. $\sqrt[3]{-64x^3 y^9}$

14. $\sqrt[4]{-16x^4 y^8}$

15. $\sqrt[4]{-81x^3 y^6}$

16. $\sqrt[3]{a^5 b^7}$

17. $\sqrt[3]{a^6 b^8}$

18. $\sqrt[3]{-8x^6 y^4}$

19. $\sqrt[3]{-64x^6 y^9}$

20. $\sqrt[3]{a^5 b^6 c^7}$

21. $\sqrt[3]{a^{12} b^9 c^8}$

22. $\sqrt[4]{a^{12} b^9 c^8}$

23. $\sqrt[4]{16x^7 y^5}$

24. $\sqrt[4]{64x^{12} y^8}$

1. _____
2. _____
3. _____
4. _____
5. _____
6. _____
7. _____
8. _____
9. _____
10. _____
11. _____
12. _____
13. _____
14. _____
15. _____
16. _____
17. _____
18. _____
19. _____
20. _____
21. _____
22. _____
23. _____
24. _____

Name _____ Score _____

Simplify.

1. $\sqrt{x^4 y^6 z^7}$ 2. $\sqrt{98a^4 b^6}$ 3. $\sqrt{50xy^6 z^9}$

1. _____

2. _____

3. _____

4. $\sqrt{75x^3 y^4 z^6}$ 5. $\sqrt{-16x^5}$ 6. $\sqrt{-x^3 y^6}$

4. _____

5. _____

6. _____

7. $\sqrt{x^5 y^7 z^9}$ 8. $\sqrt{x^6 y^8 z^4}$ 9. $\sqrt{x^6 z^{12}}$

7. _____

8. _____

9. _____

10. $x\sqrt{48xy} - \sqrt{12x^3 y}$ 11. $3\sqrt[3]{2a^5} - 5a\sqrt[3]{16a^2}$

10. _____

11. _____

12. $3a\sqrt{18a^3 b^5} + 5b\sqrt{2a^5 b^3}$ 13. $5b\sqrt{a^3 b^7} - 2a\sqrt{ab^9}$

12. _____

13. _____

14. $\sqrt[3]{8} - \sqrt[3]{27}$ 15. $\sqrt[3]{24} + \sqrt[3]{192}$

14. _____

15. _____

16. $3\sqrt[3]{2a^4} - 4a\sqrt[3]{54a}$ 17. $4y\sqrt{12y} + 3\sqrt{48y^3}$

16. _____

17. _____

18. $\sqrt{4a^3} - \sqrt{16a^3} + \sqrt{64a^3}$ 19. $3b\sqrt[4]{81a^5 b} - 2a\sqrt[4]{256ab^5}$

18. _____

19. _____

20. $\sqrt[3]{54x^4 y^3} + y\sqrt[3]{250x^4} - xy\sqrt[3]{16x}$ 21. $3a\sqrt[4]{32a^5} - 5a^2\sqrt[4]{128a} + \sqrt[4]{2a^9}$

20. _____

21. _____

22. $5y\sqrt[4]{3x^9 y} - 3xy\sqrt[4]{48x^5 y} - 2\sqrt[4]{243x^9 y^5}$

22. _____

Name _____   Score _____

Simplify.

1. $\sqrt{2}\sqrt{8}$

2. $\sqrt{6}\sqrt{8}$

3. $\sqrt{7}\sqrt{28}$

4. $\sqrt[3]{8}\sqrt[3]{32}$

5. $\sqrt{x^3 y^5}\sqrt{xy^3}$

6. $\sqrt[3]{9x^4 y^2}\sqrt[3]{6x^2 y^7}$

7. $\sqrt[3]{8a^3 b^4}\sqrt[3]{16ab^5}$

8. $\sqrt{2}\left(\sqrt{50}-\sqrt{2}\right)$

9. $\sqrt{8}\left(\sqrt{2}-\sqrt{8}\right)$

10. $\sqrt{x}\left(\sqrt{x}+\sqrt{3}\right)$

11. $\sqrt{x}\left(\sqrt{x}-\sqrt{2}\right)$

12. $\sqrt{4x}\left(\sqrt{32x}-\sqrt{18}\right)$

13. $\left(\sqrt{2x}+3\right)^2$

14. $\left(\sqrt{x}+2\right)^2$

15. $\left(\sqrt{3x}-6\right)^2$

16. $3\sqrt{2x^2}\cdot 4\sqrt{8xy^2}\cdot\sqrt{3x^3 y^4}$

17. $3\sqrt{15x^2 y}\cdot 4\sqrt{5xy^2}\cdot 5\sqrt{3x^3 y^5}$

18. $\sqrt[3]{9a^2 b}\sqrt[3]{8ab}\sqrt[3]{3a^3 b^4}$

19. $\sqrt[3]{3ab^2}\sqrt[3]{6a^2 b}\sqrt[3]{9a^5 b^6}$

20. $\left(\sqrt{2}-1\right)\left(\sqrt{2}+2\right)$

21. $\left(\sqrt{3}-3\right)\left(\sqrt{3}+1\right)$

22. $\left(\sqrt{x}-3\right)\left(\sqrt{x}+3\right)$

23. $\left(\sqrt{x}-4\right)\left(\sqrt{x}+4\right)$

1. _____
2. _____
3. _____
4. _____
5. _____
6. _____
7. _____
8. _____
9. _____
10. _____
11. _____
12. _____
13. _____
14. _____
15. _____
16. _____
17. _____
18. _____
19. _____
20. _____
21. _____
22. _____
23. _____

Name _____ Score _____

Simplify.

1. $\dfrac{\sqrt{48x^5}}{\sqrt{3x^3}}$

2. $\dfrac{\sqrt{24a^5b^3}}{\sqrt{12ab}}$

3. $\dfrac{\sqrt{27a^3b^5}}{\sqrt{3ab}}$

4. $\dfrac{1}{\sqrt{3}}$

5. $\dfrac{1}{\sqrt{5}}$

6. $\dfrac{1}{\sqrt{6x}}$

7. $\dfrac{3}{\sqrt{2y}}$

8. $\dfrac{9}{\sqrt{9x}}$

9. $\dfrac{4}{\sqrt{6a}}$

10. $\sqrt{\dfrac{5}{x}}$

11. $\sqrt{\dfrac{a}{2}}$

12. $\dfrac{2}{\sqrt[3]{3}}$

13. $\dfrac{3}{\sqrt[3]{9}}$

14. $\dfrac{\sqrt{20x^5y}}{\sqrt{40x^2y^2}}$

15. $\dfrac{\sqrt{18ab^3}}{\sqrt{48a^5b}}$

16. $\dfrac{\sqrt{24x^7y}}{\sqrt{40xy}}$

17. $\dfrac{3}{\sqrt{5}+1}$

18. $\dfrac{3}{3-\sqrt{5}}$

19. $\dfrac{2}{\sqrt{x}-4}$

20. $\dfrac{-5}{\sqrt{x}-5}$

21. $\dfrac{\sqrt{5}-\sqrt{2}}{\sqrt{5}+\sqrt{2}}$

22. $\dfrac{3\sqrt{32}-\sqrt{2}}{\sqrt{2}}$

23. $\dfrac{5\sqrt{125}+3\sqrt{80}}{\sqrt{5}}$

24. $\dfrac{\sqrt{x^3}+\sqrt{x}}{\sqrt{x}}$

25. $\dfrac{\sqrt{x^7}+\sqrt{x^5}}{\sqrt{x^5}}$

26. $\dfrac{9\sqrt{3x^3}-12x\sqrt{6x}}{\sqrt{6x}}$

27. $\dfrac{\sqrt{8x^3}-\sqrt{32x}}{\sqrt{2x}}$

1. _____
2. _____
3. _____
4. _____
5. _____
6. _____
7. _____
8. _____
9. _____
10. _____
11. _____
12. _____
13. _____
14. _____
15. _____
16. _____
17. _____
18. _____
19. _____
20. _____
21. _____
22. _____
23. _____
24. _____
25. _____
26. _____
27. _____

Name   Score

Simplify.

1. $\sqrt{x} = 3$ 2. $\sqrt{x} = 4$ 3. $\sqrt{a} = 7$

4. $\sqrt{x} = 6$ 5. $\sqrt[3]{a} = 2$ 6. $\sqrt[3]{x} = 4$

7. $\sqrt[3]{x} = 6$ 8. $\sqrt{3y} = 12$ 9. $\sqrt{2x} = 8$

10. $\sqrt{4x} = 4$ 11. $\sqrt[3]{3x} = -6$ 12. $\sqrt{3x} = -6$

13. $\sqrt{3x+1} - 7 = 0$ 14. $\sqrt{5x+1} - 4 = 0$ 15. $\sqrt{4x-3} = 5$

16. $\sqrt{3x+5} = \sqrt{4x-1}$ 17. $\sqrt{2x+3} = \sqrt{5x-9}$

18. $\sqrt[3]{3x+2} = 2$ 19. $\sqrt[3]{3x+4} = 4$

20. $\sqrt[3]{2x-6} = \sqrt[3]{x+2}$ 21. $\sqrt[3]{x-10} = \sqrt[3]{5x+18}$

22. $\sqrt[4]{3x-5} = 2$ 23. $\sqrt{3x-2} + 3 = 5$

24. $\sqrt[3]{x-3} + 6 = 4$ 25. $\sqrt[3]{2x-1} + 4 = 3$

1. _____
2. _____
3. _____
4. _____
5. _____
6. _____
7. _____
8. _____
9. _____
10. _____
11. _____
12. _____
13. _____
14. _____
15. _____
16. _____
17. _____
18. _____
19. _____
20. _____
21. _____
22. _____
23. _____
24. _____
25. _____

Name   Score
_____

Solve.

1. Find the length of a rectangle that has a diagonal of 13 ft and a width of 5 ft.

2. A 20-foot ladder is leaning against a building. How high on the building will the ladder reach when the bottom of the ladder is 7 ft from the building? Round to the nearest tenth.

1. _____

2. _____

3. How far would a submarine periscope have to be above the water to locate a ship 4.9 mi away? The equation for the distance in miles that the lookout can see is $d = 1.4\sqrt{h}$, where h is the height in feet above the surface of the water.

4. How far would a submarine periscope have to be above the water to locate a ship 5.4 mi away? The equation for the distance in miles that the lookout can see is $d = 1.4\sqrt{h}$, where h is the height in feet above the surface of the water. Round to the nearest hundredth.

3. _____

4. _____

5. An object is dropped from a tall tower. Find the distance the object has fallen when the speed reaches 75 ft/s. Use the equation $v = \sqrt{64d}$, where v is the speed of the object and d is the distance. Round to the nearest hundredth.

6. An object is dropped from a tall tower. Find the distance the object has fallen when the speed reaches 320 ft/s. Use the equation $v = \sqrt{64d}$, where v is the speed of the object and d is the distance.

5. _____

6. _____

7. Find the distance required for a car to reach a velocity of 90 ft/s when the acceleration is 25 ft/s². Use the equation $v = \sqrt{2as}$, where v is the velocity, a is the acceleration, and s is the distance.

8. Find the distance required for a car to reach a velocity of 80 ft/s when the acceleration is 16 ft/s². Use the equation $v = \sqrt{2as}$, where v is the velocity, a is the acceleration, and s is the distance.

7. _____

8. _____

9. Find the length of a pendulum that makes one swing in 2 s. The equation for the time of one swing of a pendulum is given by $T = 2\pi\sqrt{\dfrac{L}{32}}$, where T is the time in seconds and L is the length in feet. Use 3.14 for π. Round to the nearest hundredth.

10. Find the length of a pendulum that makes one swing in 3.2 s. The equation for the time of one swing of a pendulum is given by $T = 2\pi\sqrt{\dfrac{L}{32}}$, where T is the time in seconds and L is the length in feet. Use 3.14 for π. Round to the nearest hundredth.

9. _____

10. _____

Name   Score

Simplify.

1. $\sqrt{-49}$ 2. $\sqrt{-8}$ 1. _____

                                                                                        2. _____

3. $\sqrt{-12}$ 4. $\sqrt{-200}$ 3. _____

                                                                                        4. _____

5. $\sqrt{-128}$ 6. $\sqrt{-144}$ 5. _____

                                                                                        6. _____

7. $\sqrt{-108}$ 8. $\sqrt{-180}$ 7. _____

                                                                                        8. _____

9. $\sqrt{9}+\sqrt{-9}$ 10. $\sqrt{16}+\sqrt{-25}$ 9. _____

                                                                                        10. _____

11. $\sqrt{36}+\sqrt{-16}$ 12. $\sqrt{49}+\sqrt{-100}$ 11. _____

                                                                                        12. _____

13. $\sqrt{48}-\sqrt{-27}$ 14. $\sqrt{50}+\sqrt{-18}$ 13. _____

                                                                                        14. _____

15. $\sqrt{24}-\sqrt{-36}$ 16. $\sqrt{50}-\sqrt{-98}$ 15. _____

                                                                                        16. _____

17. $\sqrt{150}-\sqrt{-27}$ 18. $\sqrt{72}-\sqrt{-128}$ 17. _____

                                                                                        18. _____

Name _____ Score _____

Simplify.

1. $(5-8i)+(6+2i)$

2. $(4-6i)+(7-3i)$

3. $(4+6i)+(8-3i)$

4. $(7-4i)+(5+3i)$

5. $(4-6i)+(9-i)$

6. $(4-4i)-(10-2i)$

7. $(6-10i)-(12-11i)$

8. $(8-27i)-(7+48i)$

9. $(32+8i)+(50-72i)$

10. $(24-20i)-(96+45i)$

11. $(9-2i)+i$

12. $(8-5i)+2i$

13. $(6-4i)+3i$

14. $(7-6i)+5i$

15. $(5+4i)+(-5-4i)$

16. $(7-5i)+(-10+2i)$

17. $(8+3i)+(-6-4i)$

18. $(9-2i)+(-4+4i)$

1. _____

2. _____

3. _____

4. _____

5. _____

6. _____

7. _____

8. _____

9. _____

10. _____

11. _____

12. _____

13. _____

14. _____

15. _____

16. _____

17. _____

18. _____

Name   Score

Simplify.

1. $(5i)(-8i)$ **2.** $(-5i)(-3i)$ **3.** $(-8i)(20)$ **1.** _____

                                                                        2. _____

                                                                        3. _____

4. $(-4i)(-8i)$ **5.** $\sqrt{-3}\sqrt{-27}$ **6.** $\sqrt{-2}\sqrt{-32}$ **4.** _____

                                                                        5. _____

                                                                        6. _____

7. $\sqrt{-6}\sqrt{-8}$ **8.** $3i(5+i)$ **9.** $4i(2+3i)$ **7.** _____

                                                                        8. _____

                                                                        9. _____

10. $(5-2i)(3+i)$ **11.** $(4+3i)(3+i)$ **12.** $(5-i)^2$ **10.** _____

                                                                        11. _____

                                                                        12. _____

13. $(5-i)(2-i)$ **14.** $(3-4i)(2-i)$ **15.** $(6+2i)(6-2i)$ **13.** _____

                                                                        14. _____

                                                                        15. _____

16. $(3-5i)(3+5i)$ **17.** $(5-3i)(2+i)$ **18.** $(3-6i)(3-i)$ **16.** _____

                                                                        17. _____

                                                                        18. _____

19. $(5+4i)(2+i)$ **20.** $(5-3i)(2+i)$ **21.** $(3-6i)(3-i)$ **19.** _____

                                                                        20. _____

                                                                        21. _____

22. $(5+4i)(2+i)$ **23.** $(3-6i)(2+5i)$ **24.** $(3-i)\left(\dfrac{1}{3}+\dfrac{2}{3}i\right)$ **22.** _____

                                                                        23. _____

                                                                        24. _____

Name _____   Score _____

Simplify.

1. $\dfrac{4}{i}$

2. $\dfrac{2}{3i}$

3. $\dfrac{3}{4i}$

4. $\dfrac{4-6i}{-2i}$

5. $\dfrac{15+2i}{-5i}$

6. $\dfrac{5}{4+i}$

7. $\dfrac{8}{3+i}$

8. $\dfrac{5}{5+2i}$

9. $\dfrac{4}{4-i}$

10. $\dfrac{6}{5-i}$

11. $\dfrac{1-5i}{5+i}$

12. $\dfrac{1-4i}{4+i}$

13. $\dfrac{3+6i}{4+i}$

14. $\dfrac{4+20i}{5+i}$

15. $\dfrac{6-3i}{3-i}$

16. $\dfrac{4+8i}{2-i}$

17. $\dfrac{12}{3-6i}$

18. $\dfrac{2i}{8-2i}$

19. $\dfrac{3i}{12-6i}$

20. $\dfrac{4-5i}{5+i}$

21. $\dfrac{3-2i}{2+i}$

22. $\dfrac{4+3i}{2-i}$

23. $\dfrac{2+5i}{1-i}$

24. $\dfrac{6+5i}{2-i}$

1. _____
2. _____
3. _____
4. _____
5. _____
6. _____
7. _____
8. _____
9. _____
10. _____
11. _____
12. _____
13. _____
14. _____
15. _____
16. _____
17. _____
18. _____
19. _____
20. _____
21. _____
22. _____
23. _____
24. _____

Name   Score

Solve by factoring.

1. $y^2 + 5y = 0$

2. $t^2 - 16 = 0$

3. $p^2 - 64 = 0$

4. $s^2 + s - 12 = 0$

5. $y^2 - 4y + 4 = 0$

6. $x^2 + 8x + 16 = 0$

7. $8z^2 - 24z = 0$

8. $3y^2 + 15y = 0$

9. $t^2 - 2t = 15$

10. $p^2 + 3p = 4$

11. $v^2 + 12 = 8v$

12. $t^2 - 12 = 11t$

13. $3x^2 + x - 2 = 0$

14. $2x^2 - 5x - 3 = 0$

15. $3x^2 - 7x - 6 = 0$

16. $t + 18 = t(t + 8)$

17. $x + 12 = x(x - 3)$

18. $t + 3 = t(t + 3)$

19. $(2v - 3)(3v + 1) = 5v^2 - 9v + 12$

20. $u^2 - 3u + 9 = (u + 1)(2u + 3)$

Solve for x by factoring.

21. $x^2 + 4bx - 21b^2 = 0$

22. $x^2 - 5bx + 4b^2 = 0$

23. $x^2 - 7cx - 8c^2 = 0$

24. $x^2 - ax - 12a^2 = 0$

25. $2x^2 - 5cx + 2c^2 = 0$

26. $3x^2 + 4ax - 6a^2 = 0$

1. _____
2. _____
3. _____
4. _____
5. _____
6. _____
7. _____
8. _____
9. _____
10. _____
11. _____
12. _____
13. _____
14. _____
15. _____
16. _____
17. _____
18. _____
19. _____
20. _____
21. _____
22. _____
23. _____
24. _____
25. _____
26. _____

Name   Score

Solve by taking square roots.

1. $y^2 = 25$

2. $x^2 = 36$

3. $z^2 = -9$

4. $v^2 = -49$

5. $s^2 - 1 = 0$

6. $t^2 - 9 = 0$

7. $z^2 + 25 = 0$

8. $u^2 - 27 = 0$

9. $t^2 + 50 = 0$

10. $2(x+3)^2 = 32$

11. $(x-2)^2 = 16$

12. $6(y+1)^2 = 24$

13. $2(x-2)^2 = 162$

14. $4(z-1)^2 = 100$

15. $\left(v - \dfrac{1}{3}\right)^2 = \dfrac{1}{9}$

16. $\left(r + \dfrac{1}{2}\right)^2 = \dfrac{1}{4}$

17. $(v-5)^2 + 15 = 0$

18. $(x+4)^2 + 28 = 0$

19. $(x-1)^2 - 18 = 0$

20. $(x+2)^2 + 16 = 0$

21. $(x-3)^2 - 50 = 0$

22. $\left(x - \dfrac{3}{5}\right)^2 - 75 = 0$

23. $\left(u - \dfrac{1}{3}\right)^2 - 27 = 0$

24. $\left(x + \dfrac{1}{2}\right)^2 - 45 = 0$

1. _____
2. _____
3. _____
4. _____
5. _____
6. _____
7. _____
8. _____
9. _____
10. _____
11. _____
12. _____
13. _____
14. _____
15. _____
16. _____
17. _____
18. _____
19. _____
20. _____
21. _____
22. _____
23. _____
24. _____

Name   Score

Solve by completing the square.

1. $v^2 + 6v - 7 = 0$

2. $w^2 - 2w - 16 = 0$

3. $z^2 - 4z + 4 = 0$

4. $u^2 + 8u + 16 = 0$

5. $x^2 - 6x + 5 = 0$

6. $y^2 + 8y + 10 = 0$

7. $z^2 - 2z - 3 = 0$

8. $x^2 + 3x - 10 = 0$

9. $v^2 - 5v - 36 = 0$

10. $y^2 - 7y + 12 = 0$

11. $y^2 - 9y + 18 = 0$

12. $x^2 + 4x - 16 = 0$

13. $x^2 - 6x + 2 = 0$

14. $3x^2 + 2x - 1 = 0$

15. $2y^2 - 3y + 1 = 0$

16. $2x^2 - 5x - 12 = 0$

17. $x^2 = 10x + 24$

18. $x^2 = 6x + 16$

19. $z^2 = 4z + 5$

20. $10x^2 - 9x + 2 = 0$

21. $4x^2 - 8x + 3 = 0$

22. $3x^2 + 14x + 12 = 0$

23. $6x^2 + 7x - 3 = 0$

24. $4x^2 - 12x - 7 = 0$

25. $2x^2 = 10 - x$

26. $(x+3)(x-1) = x - 2$

27. $(y-2)^2 = 3y - 6$

1. _____
2. _____
3. _____
4. _____
5. _____
6. _____
7. _____
8. _____
9. _____
10. _____
11. _____
12. _____
13. _____
14. _____
15. _____
16. _____
17. _____
18. _____
19. _____
20. _____
21. _____
22. _____
23. _____
24. _____
25. _____
26. _____
27. _____

Name _____   Score _____

Solve by using the quadratic formula.

1. $w^2 = 5w + 50$

2. $t^2 = 3t + 40$

3. $x^2 = 9 - 8x$

4. $2x^2 - 5x + 6 = 0$

5. $8x^2 + 10x = 3$

6. $6x^2 + 7x - 3 = 0$

7. $2x^2 - 3x + 1 = 0$

8. $4x^2 + 20x + 9 = 0$

9. $2x^2 + 2x + 5 = 0$

10. $t^2 - 4t + 6 = 0$

11. $3x^2 - 6x - 1 = 0$

12. $12x^2 - 5x - 3 = 0$

13. $2x^2 - 11x + 5 = 0$

14. $6x^2 - 11x + 9 = 0$

15. $3x^2 - 3x + 4 = 0$

16. $5y(y + 3) = 4y + 12$

17. $(2x - 3)(x - 1) = 3$

18. $(3t + 1)(t - 2) = 10$

1. _____

2. _____

3. _____

4. _____

5. _____

6. _____

7. _____

8. _____

9. _____

10. _____

11. _____

12. _____

13. _____

14. _____

15. _____

16. _____

17. _____

18. _____

Solve by using the quadratic formula. Approximate solutions to the nearest thousandth.

19. $p^2 - 6p + 2 = 0$

20. $w^2 - 8w + 3 = 0$

21. $r^2 - 7r + 4 = 0$

19. _____

20. _____

21. _____

Use the discriminant to determine whether the quadratic equation has one real number solution, two real number solutions, or two complex number solutions.

22. $9x^2 + 30x + 25 = 0$

23. $2p^2 + 7p + 3 = 0$

24. $3z^2 + 1 = 0$

22. _____

23. _____

24. _____

Name   Score

Solve.

1. $z^4 - 5z^2 + 4 = 0$ **2.** $x^4 - 20x^2 + 64 = 0$ **3.** $y^4 - 17y^2 + 16 = 0$

4. $x^4 - 29x^2 + 100 = 0$ **5.** $x^4 - 8x^2 - 9 = 0$ **6.** $x^4 - 20x^2 + 64 = 0$

7. $x - x^{1/2} - 20 = 0$ **8.** $w - 20w^{1/2} - 8 = 0$ **9.** $w - 5w^{1/2} + 4 = 0$

10. $x^4 + 4x^2 - 32 = 0$ **11.** $x^4 - 7x^2 + 12 = 0$ **12.** $x^4 + x^2 - 12 = 0$

13. $x^4 - 6x^2 + 8 = 0$ **14.** $x^4 - 1 = 0$ **15.** $x^4 - 16 = 0$

16. $x^4 - 64 = 0$ **17.** $x^4 - 16x^2 + 63 = 0$ **18.** $x - 5x^{1/2} + 6 = 0$

19. $x - 9x^{1/2} + 14 = 0$ **20.** $x - 9x^{1/2} + 8 = 0$ **21.** $y^{2/3} - 5y^{1/3} + 4 = 0$

22. $4y^4 - 17y^2 + 4 = 0$ **23.** $16w^4 - 8w^2 + 1 = 0$ **24.** $z^{2/3} - 1 = 0$

1.	_____
2.	_____
3.	_____
4.	_____
5.	_____
6.	_____
7.	_____
8.	_____
9.	_____
10.	_____
11.	_____
12.	_____
13.	_____
14.	_____
15.	_____
16.	_____
17.	_____
18.	_____
19.	_____
20.	_____
21.	_____
22.	_____
23.	_____
24.	_____

Name _____   Score _____

Solve.

1. $x = \sqrt{x} + 2$

2. $\sqrt{3x+1} = x - 1$

3. $\sqrt{2s-1} = s - 2$

4. $\sqrt{3x+1} = x - 9$

5. $\sqrt{5x-4} = x$

6. $\sqrt{5x+4} = x - 2$

7. $\sqrt{2y-6} = 3 - y$

8. $\sqrt{x-2} = 4 - x$

9. $x - 1 = \sqrt{x+5}$

10. $\sqrt{x-7} = \sqrt{x} - 1$

11. $x - 3 = \sqrt{x-3}$

12. $x - 2 = \sqrt{x+4}$

13. $\sqrt{3x+1} = x - 3$

14. $\sqrt{2x-5} + 2 = x$

15. $\sqrt{x+2} = x - 4$

16. $\sqrt{x+7} + 1 = \sqrt{x+12}$

17. $\sqrt{4-x} + \sqrt{x+6} = 4$

18. $\sqrt{x+5} + \sqrt{x} = 5$

19. $\sqrt{y} + 3 = \sqrt{y+27}$

20. $\sqrt{x+2} - \sqrt{x-1} = 1$

21. $\sqrt{2x+4} = \sqrt{x} - 2$

22. $\sqrt{3x-2} - \sqrt{2x-3} = 1$

23. $\sqrt{3-2x} + \sqrt{2x+7} = 4$

1. _____
2. _____
3. _____
4. _____
5. _____
6. _____
7. _____
8. _____
9. _____
10. _____
11. _____
12. _____
13. _____
14. _____
15. _____
16. _____
17. _____
18. _____
19. _____
20. _____
21. _____
22. _____
23. _____

Name   Score

Solve.

1. $z = \dfrac{6}{z-5}$

2. $\dfrac{z}{3z-4} = \dfrac{3}{z+2}$

3. $\dfrac{3x}{x+2} = \dfrac{6}{x-1}$

4. $\dfrac{4x}{x-2} = \dfrac{12}{x-4}$

5. $\dfrac{x-3}{x+4} + x = 3$

6. $\dfrac{4v}{v-3} = \dfrac{7}{v+8}$

7. $\dfrac{x+3}{x-1} = \dfrac{x-1}{x+1}$

8. $\dfrac{12}{x-1} + \dfrac{12}{x+1} = 5$

9. $\dfrac{3x-2}{x-3} + x = 14$

10. $\dfrac{4x-1}{x-4} - x = 2$

11. $\dfrac{3v+4}{v+5} - 4v = 8$

12. $\dfrac{2x+1}{x-1} - x = -1$

13. $\dfrac{8}{3x+2} + \dfrac{2}{x} = -3$

14. $\dfrac{14}{z-3} + \dfrac{3}{z-3} = -5$

15. $\dfrac{5x-3}{x+6} - \dfrac{2x+1}{x+4} = -1$

16. $\dfrac{3x-1}{x+5} + \dfrac{2x-3}{x+2} = 4$

17. $\dfrac{2w}{w+3} + \dfrac{1}{w+1} = 2$

18. $\dfrac{3v}{v+3} + \dfrac{4}{v+5} = 2$

19. $\dfrac{x+4}{x+2} - \dfrac{x-1}{x+3} = 4$

1. _____

2. _____

3. _____

4. _____

5. _____

6. _____

7. _____

8. _____

9. _____

10. _____

11. _____

12. _____

13. _____

14. _____

15. _____

16. _____

17. _____

16. _____

17. _____

Name   Score

Solve.

1. The length of a rectangle is 2 ft less than three times the width. The area of the rectangle is 40 ft². Find the length and width of the rectangle.

2. The height of a triangle is 3 m less than the length of the base. The area of the triangle is 20 m². Find the height of the triangle and the length of its base.

1. _____

2. _____

3. The difference between the squares of two consecutive integers is eleven. Find the two integers.

4. The sum of the squares of two consecutive odd integers is thirty-four. Find the two integers.

3. _____

4. _____

5. Five times an integer plus three times the square of the integer is twenty-two. Find the integer.

6. The difference between the squares of two consecutive odd integers is thirty-two. Find the two integers.

5. _____

6. _____

7. An old pump requires 9 h longer to empty a pool than does a new pump. With both pumps working, the pool can be emptied in 6 h. Find the time required for the new pump, working alone, to empty the pool.

8. A small air conditioner requires 16 min longer to cool a room 4° than does a larger air conditioner. Working together, the two air conditioners can cool the room 4° in 6 min. How long would it take each air conditioner, working alone, to cool the room 4°?

7. _____

8. _____

9. An old copier takes 15 min longer to copy a report than does a second, new model. With both copiers working, the report can be printed in 18 min. How long would it take each copier, working alone, to copy the report?

10. A car travels 240 mi. A second car, traveling 12 mph faster than the first, makes the same trip in 1 h less time. Find the speed of each car.

9. _____

10. _____

Name   Score

Solve and graph the solution set.

1. $(x-3)(x+3) > 0$

-5 -4 -3 -2 -1 0 1 2 3 4 5

2. $(x+4)(x+1) \geq 0$

-5 -4 -3 -2 -1 0 1 2 3 4 5

1. _____

2. _____

3. $x^2 - x - 6 < 0$

-5 -4 -3 -2 -1 0 1 2 3 4 5

4. $x^2 + 4x - 5 \leq 0$

-5 -4 -3 -2 -1 0 1 2 3 4 5

3. _____

4. _____

5. $(x-4)(x+4)(x+3) < 0$

-5 -4 -3 -2 -1 0 1 2 3 4 5

6. $(x+5)(x+1)(x-3) > 0$

-5 -4 -3 -2 -1 0 1 2 3 4 5

5. _____

6. _____

7. $\dfrac{x+5}{x-1} > 0$

-5 -4 -3 -2 -1 0 1 2 3 4 5

8. $(x+3)(x-3)(x-2) \geq 0$

-5 -4 -3 -2 -1 0 1 2 3 4 5

7. _____

8. _____

9. $(x-2)(x+4)(x-1) \leq 0$

-5 -4 -3 -2 -1 0 1 2 3 4 5

10. $\dfrac{x+3}{x} > 0$

-5 -4 -3 -2 -1 0 1 2 3 4 5

9. _____

10. _____

11. $x^2 \leq 16$

-5 -4 -3 -2 -1 0 1 2 3 4 5

12. $\dfrac{2x}{x-3} > 1$

-5 -4 -3 -2 -1 0 1 2 3 4 5

11. _____

12. _____

Name　　　　　　　　　　　　　　　　　　　　　　　Score

Graph.

1. $y = x^2 - 1$

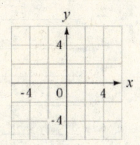

2. $y = -x^2 + 1$

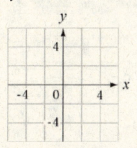

3. $y = x^2 - 3x$

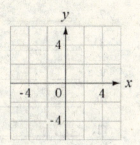

4. $y = x^2 + 3x$

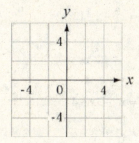

5. $y = x^2 + x - 2$

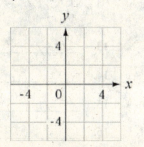

6. $y = \frac{1}{2}x^2 + 2$

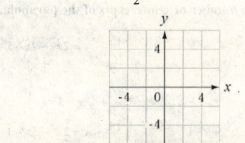

7. $y = 2x^2 + x - 3$

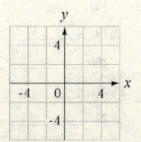

8. $y = -x^2 + 3x - 2$

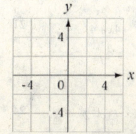

Name   Score

Find the x-intercepts of the parabola.

1. $y = x^2 - 1$
2. $y = x^2 - 16$
3. $y = x^2 - x$

4. $y = 4x^2 - 8x$
5. $y = 2x^2 - 3x - 2$
6. $y = x^2 + 5x - 6$

Find the zeros of the function.

7. $f(x) = 3x^2 - 4x + 1$
8. $f(x) = x^2 - x - 12$
9. $f(x) = x^2 + 4x + 8$

10. $f(x) = -x^2 - 3x - 4$
11. $f(x) = x^2 - 3x - 4$
12. $f(x) = x^2 + 2x + 15$

Use the discriminant to determine the number of x-intercepts of the parabola.

13. $y = 3x^2 - 5x - 2$
14. $y = x^2 + x + 2$
15. $y = 2x^2 - 9x - 5$

16. $y = 3x^2 + 2x + 2$
17. $y = 2x^2 + x - 1$
18. $y = 2x^2 + 5x - 3$

19. $y = 3x^2 + x + 2$
20. $y = x^2 - 4x + 4$
21. $y = x^2 - 6x + 9$

22. $y = 4x^2 + x - 3$
23. $y = 2x^2 - 3x - 2$
24. $y = -2x^2 + 3x - 3$

1. _____
2. _____
3. _____
4. _____
5. _____
6. _____
7. _____
8. _____
9. _____
10. _____
11. _____
12. _____
13. _____
14. _____
15. _____
16. _____
17. _____
18. _____
19. _____
20. _____
21. _____
22. _____
23. _____
24. _____

Name   Score

Find the minimum or maximum value of the quadratic function.

1. $f(x) = x^2 + 6x + 8$ 2. $f(x) = x^2 + 2x - 3$ 1. _____

                                                                     2. _____

3. $f(x) = -2x^2 + 8x + 5$ 4. $f(x) = -x^2 - 2x + 1$ 3. _____

                                                                     4. _____

5. $f(x) = x^2 + 6x - 2$ 6. $f(x) = x^2 - 3x + 2$ 5. _____

                                                                     6. _____

7. $f(x) = -3x^2 - 12x + 1$ 8. $f(x) = 2x^2 + 4x - 5$ 7. _____

                                                                     8. _____

9. $f(x) = -x^2 - x + 6$ 10. $f(x) = x^2 - 4x - 21$ 9. _____

                                                                     10. _____

11. $f(x) = 2x^2 + 6x - 1$ 12. $f(x) = -2x^2 + 4x - 1$ 11. _____

                                                                     12. _____

13. $f(x) = x^2 + 6x + 6$ 14. $f(x) = x^2 - 3x + 6$ 13. _____

                                                                     14. _____

15. $f(x) = -3x^2 + 4x + 6$ 16. $f(x) = -2x^2 - 3x + 1$ 15. _____

                                                                     16. _____

Name   Score

Solve.

1. The height in feet (s) of a rock thrown straight up is given by the function $s(t) = -16t^2 + 96t$, where t is the time in seconds. Find the maximum height above the ground that the rock will attain.

2. The height in feet (s) of a ball thrown upward at an initial speed of 72 ft/s from a platform 50 ft high is given by the function $s(t) = -16t^2 + 72t + 50$, where t is the time in seconds. Find the maximum height above the ground that the ball will attain.

1. _____

2. _____

3. A pool is treated with a chemical to reduce the amount of algae. The amount of algae in the pool t days after the treatment can be approximated by the function $A(t) = 30t^2 - 300t + 400$. How many days after treatment will the pool have the least amount of algae?

4. A manufacturer of videocassette recorders believes that the revenue the company receives is related to the price (P) of a recorder by the function $R(P) = 75P - \frac{1}{5}P^2$. What price will give the maximum revenue?

3. _____

4. _____

5. The height in feet (s) of a rocket after t seconds is given by the formula $s(t) = 224t - 16t^2$. After how many seconds does the rocket reach its maximum height?

6. Find the two numbers whose sum is 50 and whose product is a maximum.

5. _____

6. _____

7. Find two numbers whose sum is 30 and whose product is a maximum.

8. Find two numbers whose difference is 20 and whose product is a minimum.

7. _____

8. _____

9. A rectangle has a perimeter of 56 ft. Find the dimensions of the rectangle with this perimeter that will have the maximum area. What is the maximum area?

10. The perimeter of a rectangular window is 32 ft. Find the dimensions of the window with this perimeter that will enclose the largest area. What is the maximum area?

9. _____

10. _____

Name _____   Score _____

Given $f(x) = 3x^2 - 4$ **and** $g(x) = -5x + 6$, **find:**

1. $f(-2) - g(-2)$ **2.** $f(-1) + g(-1)$ **1.** _____

                                                                     2. _____

3. $(f \cdot g)(-1)$ **4.** $\left(\dfrac{g}{f}\right)(-1)$ **3.** _____

                                                                     4. _____

5. $\left(\dfrac{f}{g}\right)(-2)$ **6.** $g(5) - f(5)$ **5.** _____

                                                                     6. _____

Given $f(x) = 5x^2 - 4x - 6$ **and** $g(x) = 3x - 5$, **find:**

7. $f(5) - g(5)$ **8.** $g(1) - f(1)$ **7.** _____

                                                                     8. _____

9. $f(2) + g(2)$ **10.** $(f \cdot g)(-2)$ **9.** _____

                                                                     10. _____

11. $\left(\dfrac{f}{g}\right)\left(\dfrac{5}{3}\right)$ **12.** $\left(\dfrac{g}{f}\right)(0)$ **11.** _____

                                                                     12. _____

Given $f(x) = x^2 + 5x - 4$ **and** $g(x) = x^2 - 2x + 7$, **find:**

13. $f(-2) - g(-2)$ **14.** $f(3) + g(3)$ **13.** _____

                                                                     14. _____

15. $(f \cdot g)(1)$ **16.** $\left(\dfrac{f}{g}\right)(-3)$ **15.** _____

                                                                     16. _____

Name   Score

Given $f(x) = 5x - 4$ **and** $g(x) = 2x + 7$**, find:**

1. $f(g(1))$ **2.** $g(f(1))$ **1.** _____

                                                                         2. _____

3. $f(g(2))$ **4.** $g(f(-2))$ **3.** _____

                                                                         4. _____

5. $g(f(x))$ **6.** $f(g(x))$ **5.** _____

                                                                         6. _____

Given $h(x) = x^2 + 5$ **and** $g(x) = x - 3$**, find:**

7. $g(h(3))$ **8.** $h(g(x))$ **7.** _____

                                                                         8. _____

9. $g(h(x))$ **10.** $h(g(0))$ **9.** _____

                                                                         10. _____

Given $h(x) = 2x^3 - 3$ **and** $f(x) = 4x - 5$**, find:**

11. $f(h(-1))$ **12.** $h(f(-1))$ **11.** _____

                                                                         12. _____

13. $f(h(2))$ **14.** $f(h(x))$ **13.** _____

                                                                         14. _____

15. $h(f(x))$ **16.** $f(h(0))$ **15.** _____

                                                                         16. _____

Name   Score

Determine if the graph represents the graph of a 1-1 function.

1.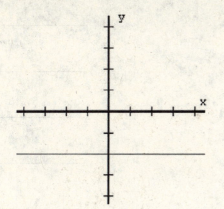

2.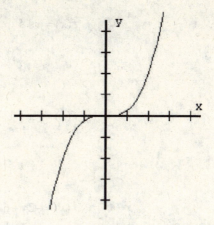

1. _____

2. _____

3.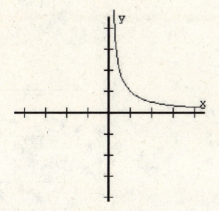

4.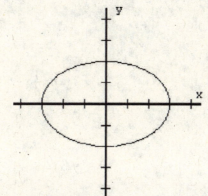

3. _____

4. _____

5.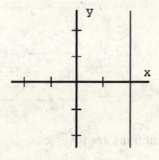

6.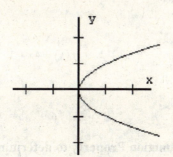

5. _____

6. _____

7.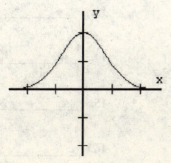

8.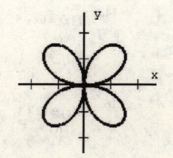

7. _____

8. _____

Name ·   Score

Find the inverse of the function.

1. $f(x) = 4x + 8$

2. $f(x) = x + 5$

1. _____

2. _____

3. $f(x) = 3x - 6$

4. $f(x) = -x + 4$

3. _____

4. _____

5. $f(x) = -\dfrac{1}{4}x + 2$

6. $f(x) = -2x + 6$

5. _____

6. _____

7. $f(x) = \dfrac{2}{3}x - 2$

8. $f(x) = \dfrac{1}{4}x - 6$

7. _____

8. _____

9. $f(x) = 3x + 2$

10. $f(x) = 5x + 1$

9. _____

10. _____

11. $f(x) = -4x + 2$

12. $f(x) = 4x + 3$

11. _____

12. _____

Use the composition of Inverse Function Property to determine whether the functions are inverses.

13. $f(x) = -\dfrac{1}{2}x$, $g(x) = -2x$

14. $f(x) = 6x + 5$, $g(x) = \dfrac{1}{6}x - \dfrac{5}{6}$

13. _____

14. _____

15. $g(x) = x + 5$, $h(x) = 5 - x$

16. $h(x) = 7x + 3$, $g(x) = \dfrac{1}{7}x - 3$

15. _____

16. _____

Name _____   Score _____

Given the function $f(x) = 4^x$ **, find:**

1. $f(1)$ 2. $f(-1)$ 3. $f(0)$

4. $f(2)$ 5. $f(-2)$ 6. $f(3)$

Given the function $f(x) = 3^{x+1}$ **, find:**

7. $f(0)$ 8. $f(1)$ 9. $f(-2)$

Given the function $f(x) = \left(\frac{1}{3}\right)^{2x}$ **, find:**

10. $f(0)$ 11. $f(1)$ 12. $f(-1)$

13. $f\left(\frac{1}{2}\right)$ 14. $f\left(-\frac{1}{2}\right)$ 15. $f(2)$

Given the function $f(x) = 3^{x^2}$ **, find:**

16. $f(0)$ 17. $f(1)$ 18. $f(2)$

19. $f(-1)$ 20. $f(-2)$ 21. $f(3)$

1. _____
2. _____
3. _____
4. _____
5. _____
6. _____
7. _____
8. _____
9. _____
10. _____
11. _____
12. _____
13. _____
14. _____
15. _____
16. _____
17. _____
18. _____
19. _____
20. _____
21. _____

Name   Score

Graph.

1. $f(x) = 2^x$

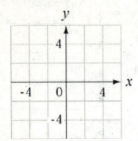

2. $f(x) = 2^{-x}$

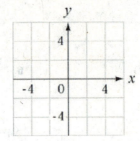

3. $f(x) = 3^{x+2}$

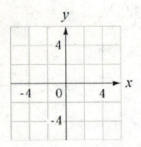

4. $f(x) = \left(\dfrac{1}{3}\right)^{-x-1}$

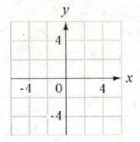

5. $f(x) = \left(\dfrac{1}{2}\right)^x$

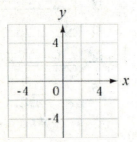

6. $f(x) = \left(\dfrac{1}{2}\right)^x - 1$

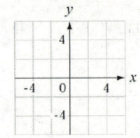

7. $f(x) = 2^x - 3$

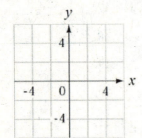

8. $f(x) = 2^x + 1$

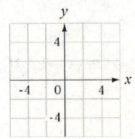

Name   Score

Write the exponential expression in logarithmic form.

1. $4^3 = 64$

2. $7^2 = 49$

3. $64^{-2} = \frac{1}{8}$

1. _____

2. _____

3. _____

4. $10^4 = 10,000$

5. $4^0 = 1$

6. $p^x = q$

4. _____

5. _____

6. _____

Write the logarithmic expression in exponential form.

7. $\log_2 1 = 0$

8. $\log_2 \left(\frac{1}{4} \right) = -2$

9. $\ln x = c$

7. _____

8. _____

9. _____

Evaluate.

10. $\log_5 25$

11. $\log 0.0001$

12. $\log_4 16$

10. _____

11. _____

12. _____

13. $\ln e^4$

14. $\log_5 625$

15. $\log_8 1$

13. _____

14. _____

15. _____

Solve for _x_.

16. $\log_8 x = 2$

17. $\log_4 x = 2$

18. $\log_6 x = 3$

16. _____

17. _____

18. _____

19. $\log_7 x = 0$

20. $\log_{10} x = 4$

21. $\log_5 x = 3$

19. _____

20. _____

21. _____

Name   Score

Express as a single logarithm with a coefficient of 1.

1. $\log_2 x^2 + \log_2 y$

2. $\log_5 x^3 - \log_5 y^4$

3. $4\log_5 x$

4. $-3\log_6 x$

5. $3\log_6 x - 4\log_6 y$

6. $3\log_3 x - 2\log_3 y + 3\log_3 z$

7. $\log_a x - (2\log_a y - \log_a z)$

8. $4(\log_2 x + \log_2 y)$

9. $\frac{1}{2}(\log_5 x + \log_5 z)$

10. $\frac{1}{3}(\log_6 x - \log_6 z)$

11. $5\log_3 x - 3(\log_3 y - \log_3 z)$

12. $\frac{1}{2}(5\log_5 x - 3\log_5 y + \log_5 z)$

1. _____

2. _____

3. _____

4. _____

5. _____

6. _____

7. _____

8. _____

9. _____

10. _____

11. _____

12. _____

Write the logarithm in expanded form.

13. $\log_3(2x)$

14. $\log_2 x^3$

15. $\log_4\left(\dfrac{x}{y}\right)$

16. $\log_6\left(\dfrac{u^4}{v^5}\right)$

17. $\log_3(xy)^3$

18. $\ln xy^3 z^2$

19. $\log_6 \sqrt{xy}$

20. $\log_6 \sqrt[3]{r^2 w}$

21. $\log_5 \sqrt{\dfrac{x^2}{y}}$

22. $\log_5 \dfrac{\sqrt{x}}{y}$

13. _____

14. _____

15. _____

16. _____

17. _____

18. _____

19. _____

20. _____

21. _____

22. _____

Name _____   Score

Find the logarithm.

1. $\log_4 9$ 2. $\log_5 8$ 1. _____

2. _____

3. $\log_7 4$ 4. $\log_8 7$ 3. _____

4. _____

5. $\log_6 12$ 6. $\log_9 15$ 5. _____

6. _____

7. $\log_4 12$ 8. $\log_5 24$ 7. _____

8. _____

9. $\log_6 30$ 10. $\log_3 9.2$ 9. _____

10. _____

11. $\log_5 8.6$ 12. $\log_9 5$ 11. _____

12. _____

13. $\log_8 6$ 14. $\log_3 2.9$ 13. _____

14. _____

15. $\log_5 9$ 16. $\log_7 1.9$ 15. _____

16. _____

17. $\log_6 2.7$ 18. $\log_7 1.2$ 17. _____

18. _____

Name

Score

Graph.

1. $f(x) = \log_3 x$

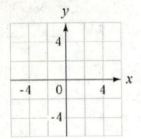

2. $f(x) = \log_2 x + 1$

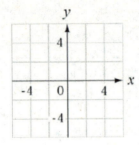

3. $f(x) = \log_2 3x$

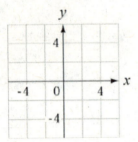

4. $f(x) = -\log_3(2 - x)$

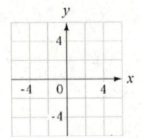

5. $f(x) = \log_3(x + 2)$

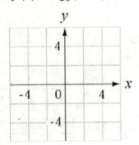

6. $f(x) = -\dfrac{1}{2}\log_2 x$

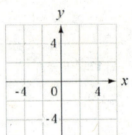

7. $f(x) = -\log_4 x$

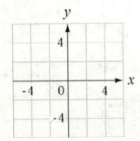

8. $f(x) = \log_4(x + 2)$

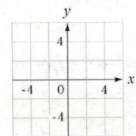

Name   Score

Solve for *x*. Round to the nearest ten-thousandth.

1. $3^{2x-1} = 3^{x-3}$

2. $2^{4x-1} = 2^{x+2}$

3. $10^{3x-4} = 10^{x+6}$

4. $8^{x-3} = 8^{3x+5}$

5. $10^x = 6$

6. $12^x = 9$

7. $\left(\dfrac{1}{2}\right)^x = 4$

8. $\left(\dfrac{1}{3}\right)^x = 81$

9. $(1.4)^x = 3$

10. $(2.5)^x = 5$

11. $10^x = 24$

12. $10^x = 38$

13. $2^{-x} = 8$

14. $3^{-x} = 15$

15. $2^{x-1} = 9$

16. $3^{x+1} = 7$

17. $4^{2x-1} = 64$

18. $3^{-x+2} = 15$

19. $5^x = 25^{2x-1}$

20. $2^{-x+6} = 32^x$

21. $4^x = 2^{x+3}$

22. $6^{x-1} = 36^x$

23. $9^{x+6} = 27^{2x}$

24. $4^{3x} = 16^{x-2}$

1. _____
2. _____
3. _____
4. _____
5. _____
6. _____
7. _____
8. _____
9. _____
10. _____
11. _____
12. _____
13. _____
14. _____
15. _____
16. _____
17. _____
18. _____
19. _____
20. _____
21. _____
22. _____
23. _____
24. _____

Name _____　　　　　　　　　Score

Solve for x.

1. $\log_5(x+1)=2$

2. $\log_7(x+2)=1$

3. $\log_3(2x-1)=3$

4. $\log_5(x-1)=2$

5. $\log_4(3x+1)=2$

6. $\log_3(2x-3)=4$

7. $\log_5\left(x^2-4x\right)=1$

8. $\log_4\left(x^2+6x\right)=2$

9. $\log_6\left(x^2-5x\right)=2$

10. $\log_3 x=\log_3(2-x)$

11. $\log_5\left(\dfrac{3x}{x-1}\right)=1$

12. $\log_4 x+\log_4(3x-4)=\log_4 15$

13. $\log_7 5x=\log_7 2+\log_7(x-3)$

14. $\log_5 9x-\log_5\left(x^2-2\right)=\log_5 6$

15. $\log_6 3x=\log_6 2+\log_6\left(x^2-1\right)$

16. $\log_4 7x=\log_4 3+\log_4\left(x^2-2\right)$

1. _____

2. _____

3. _____

4. _____

5. _____

6. _____

7. _____

8. _____

9. _____

10. _____

11. _____

12. _____

13. _____

14. _____

15. _____

16. _____

Name _____　　Score _____

Use the compound interest formula $P = A(1+i)^n$, **where** A **is the original value of an investment,** i **is the interest rate per compounding period,** n **is the total number of compounding periods, and** P **is the value of the investment after** n **periods.**

1. An investor deposits $8000 into an account that earns 12% interest compounded semi-annually. What is the value of the investment after 4 years?

2. An investor deposits $5000 into an account that earns 12% annual interest compounded semi-annually. In approximately how many years will the investment be worth twice the original amount?

1. _____

2. _____

The percent of correct welds a student can make will increase with practice and can be approximated by the equation $P = 100[1 - (0.75)^t]$, **where** P **is the percent of correct welds and** t **is the number of weeks of practice.**

3. Find the percent of correct welds a student will make after four weeks of practice.

4. How many weeks of practice are necessary before a student will make 95% of the welds correctly?

3. _____

4. _____

The percent of light that will pass through a material is given by the equation $\log P = -kd$, **where** P **is the percent of light passing through the material,** k **is a constant that depends on the material, and** d **is the thickness of the material in centimeters.**

5. The constant k for a piece of tinted glass is 0.5. How thick a piece of this glass is needed so that 60% of the light that is incident to the glass will pass through it?

6. The constant k for a piece of opaque glass that is 0.25 cm thick is 0.5. Find the percent of light that will pass through the glass.

5. _____

6. _____

Use the Richter equation $M = \log \dfrac{I}{I_0}$, **where** M **is the magnitude of an earthquake,** I **is the intensity of its shock waves, and** I_0 **is a constant.**

7. How many times stronger is an earthquake that has magnitude 5.8 on the Richter scale than one that has magnitude 3.2 on the scale?

8. How many times stronger is an earthquake that has magnitude 6.4 on the Richter scale than one that has magnitude 1.6 on the scale?

7. _____

8. _____

125

Name   Score

Find the vertex and axis of symmetry of the parabola. Then sketch its graph.

1. $y = x^2 - x - 2$ 2. $y = x^2 + 2x - 3$ 1. _____

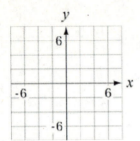   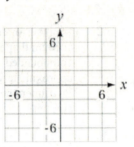

2. _____

3. $y = -x^2 + 2x + 4$ 4. $y = -x^2 + 6x - 5$ 3. _____

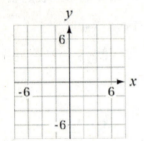   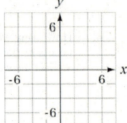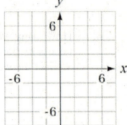

4. _____

5. $x = y^2 + 2x + 3$ 6. $x = 3y^2 - 3$ 5. _____

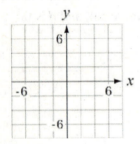   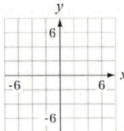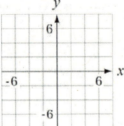

6. _____

7. $x = -\dfrac{1}{4}y^2 + 4$ 8. $x = -\dfrac{1}{2}y^2 + 2$ 7. _____

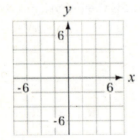   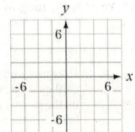

8. _____

Name   Score

Sketch a graph of the circle.

1. $(x-1)^2 + (y+1)^2 = 4$

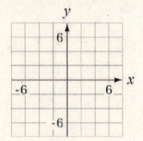

2. $(x+1)^2 + (y-2)^2 = 9$

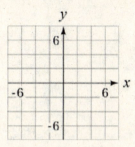

3. $(x-1)^2 + (y+3)^3 = 9$

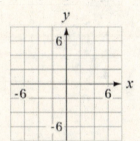

4. $(x-3)^2 + (y+2)^2 = 16$

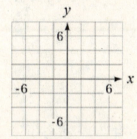

5.   Find the equation of the circle with radius 4 and center (0, –4). Then sketch the graph.

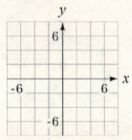

6.   Find the equation of the circle with radius 2 and center (–2, –1). Then sketch the graph.

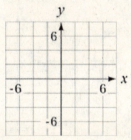

7.   Find the equation of the circle that passes through point (3, 0) and whose center is (–1, 3). Then sketch its graph.

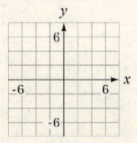

8.   Find the equation of the circle that passes through point (–2, –3) and whose center is (2, –3). Then sketch its graph.

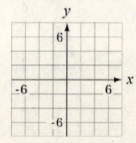

Name   Score

Write the equation of the circle in standard form. Then sketch its graph.

1. $x^2 - 10x + y^2 + 2y + 22 = 0$ **2.** $x^2 + y^2 + 8x - 4y + 4 = 0$ 1. _____

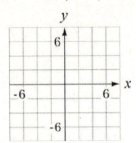   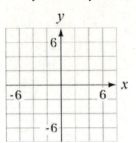

2. _____

3. $x^2 + y^2 - 10x + 5y + 25 = 0$ **4.** $x^2 + y^2 + 6x - 10y + 9 = 0$ 3. _____

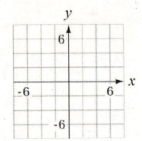   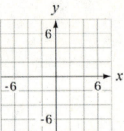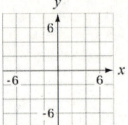

4. _____

5. $x^2 + y^2 - 6x + 8y - 16 = 0$ **6.** $x^2 + y^2 - 6x - 4y + 9 = 0$ 5. _____

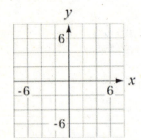   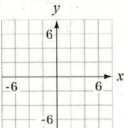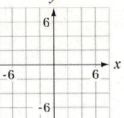

6. _____

7. $x^2 + y^2 + 6x - 4y - 3 = 0$ **8.** $x^2 + y^2 + 6x - 2y + 6 = 0$ 7. _____

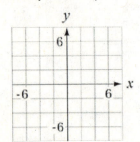   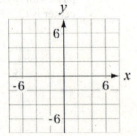

8. _____

Name

Score

Sketch a graph of the ellipse.

1.

$$\frac{x^2}{9} + \frac{y^2}{4} = 1$$

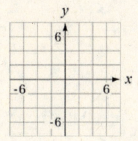

2.

$$\frac{x^2}{25} + \frac{y^2}{9} = 1$$

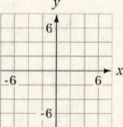

3.

$$\frac{x^2}{36} + \frac{y^2}{9} = 1$$

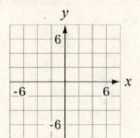

4.

$$\frac{x^2}{4} + \frac{y^2}{25} = 1$$

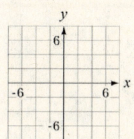

5.

$$\frac{x^2}{49} + \frac{y^2}{25} = 1$$

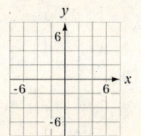

6.

$$\frac{x^2}{49} + \frac{y^2}{4} = 1$$

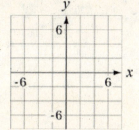

7.

$$\frac{x^2}{25} + \frac{y^2}{49} = 1$$

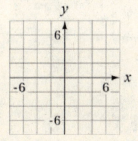

8.

$$\frac{x^2}{36} + \frac{y^2}{25} = 1$$

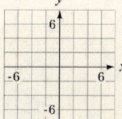

Name   Score

Sketch a graph of the hyperbola.

1.　$\dfrac{y^2}{4} - \dfrac{x^2}{9} = 1$

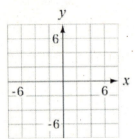

2.　$\dfrac{x^2}{9} - \dfrac{y^2}{25} = 1$

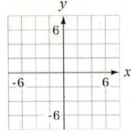

3.　$\dfrac{y^2}{49} - \dfrac{x^2}{4} = 1$

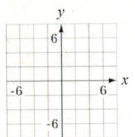

4.　$\dfrac{y^2}{9} - \dfrac{x^2}{9} = 1$

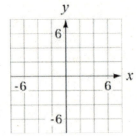

5.　$\dfrac{x^2}{4} - \dfrac{y^2}{25} = 1$

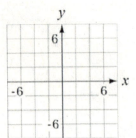

6.　$\dfrac{x^2}{49} - \dfrac{y^2}{9} = 1$

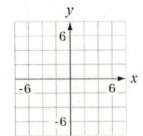

7.　$\dfrac{y^2}{36} - \dfrac{x^2}{9} = 1$

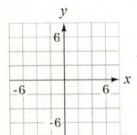

8.　$\dfrac{y^2}{9} - \dfrac{x^2}{36} = 1$

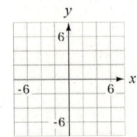

Name _____ Score _____

Solve.

1. $y = x^2 - 3x - 4$
     $y = 2x + 2$

2. $y = x^2 - 3x - 9$
     $y = 2x + 5$

1. _____

2. _____

3. $y = x^2 + 4x - 18$
     $2x - y = 3$

4. $y = x^2 - 5x$
     $y = 2x$

3. _____

4. _____

5. $y^2 = x + 2$
     $y = x$

6. $x^2 + 4y^2 = 13$
     $x + 2y = -5$

5. _____

6. _____

7. $x^2 + y^2 = 10$
     $x + y = 4$

8. $x^2 + y^2 = 25$
     $2x - y = 2$

7. _____

8. _____

9. $x^2 + y^2 = 13$
     $x - y = 1$

10. $x^2 - y^2 = 7$
      $x - 2y = -2$

9. _____

10. _____

11. $y^2 - x^2 = 5$
      $x + y = 1$

12. $x^2 + y^2 = 45$
      $x^2 - y^2 = 27$

11. _____

12. _____

13. $2x^2 + y^2 = 8$
      $x^2 - 2y^2 = 4$

14. $x^2 + 4y^2 = 25$
      $x^2 - y^2 = 5$

13. _____

14. _____

15. $2x^2 + y^2 = 6$
      $x - 2y = -5$

16. $x^2 + 4y^2 = 17$
      $3x^2 - y^2 = -1$

15. _____

16. _____

Name   Score

Graph the solution set.

1. $y < x^2 - 5x + 4$

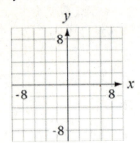

2. $y > x^2 - 2x$

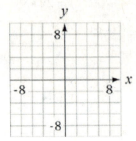

3. $(x-2)^2 + (y+1)^2 \le 4$

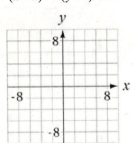

4. $(x+1)^2 + (y-3)^2 > 4$

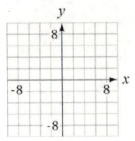

5. $\dfrac{x^2}{9} + \dfrac{y^2}{16} > 1$

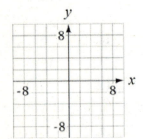

6. $\dfrac{x^2}{36} + \dfrac{y^2}{9} \le 1$

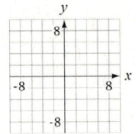

7. $\dfrac{x^2}{25} - \dfrac{y^2}{9} \ge 1$

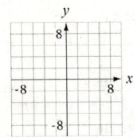

8. $\dfrac{y^2}{4} - \dfrac{x^2}{9} \le 1$

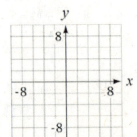

Name   Score

Graph the solution set.

1.
$$y \le (x-1)^2$$
$$x + y > 3$$

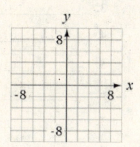

2.
$$\frac{x^2}{25} + \frac{y^2}{4} < 1$$
$$x + y \ge 8$$

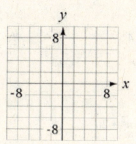

3.
$$y > x^2 - 8$$
$$y < x - 4$$

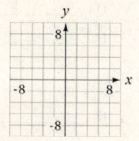

4.
$$\frac{x^2}{36} + \frac{y^2}{16} \le 1$$
$$x + 2y \le 2$$

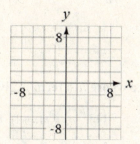

5.
$$x \ge y^2 - 4$$
$$y \le 2x - 6$$

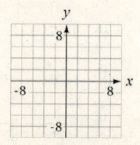

6.
$$\frac{x^2}{36} + \frac{y^2}{16} < 1$$
$$x^2 + y^2 > 4$$

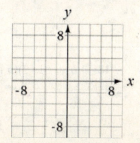

Name   Score

Write the first four terms of the sequence whose *n*th term is given by the formula.

1. $a_n = n + 2$ 2. $a_n = n - 3$ 3. $a_n = 2 - n$ 1. _____

                                                        2. _____

                                                        3. _____

4. $a_n = 1 + 2n$ 5. $a_n = 4^n$ 6. $a_n = 2^{n-1}$ 4. _____

                                                        5. _____

                                                        6. _____

7. $a_n = (n+1)(n+2)$ 8. $a_n = n^2 + 2n$ 9. $a_n = \dfrac{n^2 + 1}{n}$ 7. _____

                                                        8. _____

                                                        9. _____

10. $a_n = n + \dfrac{1}{n}$ 11. $a_n = n^2 + \dfrac{1}{n}$ 12. $a_n = (-1)^n n$ 10. _____

                                                        11. _____

                                                        12. _____

Find the indicated term of the sequence whose *n*th term is given by the formula.

13. $a_n = 5n - 12; \ a_9$ 14. $a_n = n(n+1); \ a_{12}$ 13. _____

                                                        14. _____

15. $a_n = \dfrac{n}{n-1}; \ a_{10}$ 16. $a_n = 2^n; \ a_5$ 15. _____

                                                        16. _____

17. $a_n = (n+4)(n+2); \ a_{16}$ 18. $a_n = (n+3)(n+1); \ a_6$ 17. _____

                                                        18. _____

19. $a_n = \dfrac{(-1)^{2n}}{n^2}; \ a_{10}$ 20. $a_n = \dfrac{(-2)^{n+1}}{n+1}; \ a_7$ 19. _____

                                                        20. _____

Name   Score _____

Find the sum of the series.

1. $\displaystyle\sum_{i=1}^{4} 3i$

2. $\displaystyle\sum_{n=1}^{6} (n+3)$

3. $\displaystyle\sum_{i=1}^{4} (4i+3)$

4. $\displaystyle\sum_{i=1}^{4} i^3$

5. $\displaystyle\sum_{i=1}^{5} (i^2+2)$

6. $\displaystyle\sum_{n=1}^{3} 3^n$

7. $\displaystyle\sum_{n=2}^{6} \frac{n-1}{n}$

8. $\displaystyle\sum_{i=3}^{6} (2i-3)^2$

9. $\displaystyle\sum_{i=1}^{4} \frac{1}{3i}$

10. $\displaystyle\sum_{i=3}^{6} \frac{1-i}{i}$

11. $\displaystyle\sum_{n=1}^{6} 3n^2+1$

12. $\displaystyle\sum_{n=2}^{6} (3n^2+1)$

1. _____
2. _____
3. _____
4. _____
5. _____
6. _____
7. _____
8. _____
9. _____
10. _____
11. _____
12. _____

Write the series in expanded form.

13. $\displaystyle\sum_{n=1}^{4} x^n$

14. $\displaystyle\sum_{i=1}^{4} \frac{x^i}{3i}$

15. $\displaystyle\sum_{n=1}^{5} 3x^n$

16. $\displaystyle\sum_{i=3}^{5} \frac{x^i}{i^3}$

17. $\displaystyle\sum_{n=1}^{4} x^{3n}$

18. $\displaystyle\sum_{i=3}^{6} \frac{x^i}{2i}$

19. $\displaystyle\sum_{n=1}^{4} (3n)x^n$

20. $\displaystyle\sum_{n=1}^{4} nx^{n+1}$

21. $\displaystyle\sum_{i=1}^{4} (x^{-i})^2$

13. _____
14. _____
15. _____
16. _____
17. _____
18. _____
19. _____
20. _____
21. _____

135

Name   Score

Find the indicated term of the arithmetic sequence.

1. 3, 7, 11, …; a_{10} **2.** 2, 7, 12, …; a_{15} **1.** _____

                                                                        2. _____

3. −4, −2, 0, …; a_{20} **4.** −9, −7, −5, …; a_{12} **3.** _____

                                                                        4. _____

5. 4, 8, 12, …; a_{25} **6.** −3, 2, 7, …; a_{15} **5.** _____

                                                                        6. _____

Find the formula for the *n*th term of the arithmetic sequence.

7. 1, 5, 9, …   **8.** 1, 6, 11, …   **9.** 9, 7, 5, …   **7.** _____

                                                                        8. _____

                                                                        9. _____

10. 4, 7, 10, …   **11.** $1, \dfrac{7}{2}, 6, \ldots$ **12.** 5, −3, −11, …   **10.** _____

                                                                        11. _____

                                                                        12. _____

13. −8, −10, −12, …   **14.** 8, 13, 18, …   **15.** 13, 7, 1, …   **13.** _____

                                                                        14. _____

                                                                        15. _____

Find the number of terms in the finite arithmetic sequence.

16. 1, 3, 5, …, 15   **17.** 2, 6, 10, … 78   **16.** _____

                                                                        17. _____

18. 2, −2, −6, …, −66   **19.** −6, −10, −14, …, −46   **18.** _____

                                                                        19. _____

20. $\dfrac{6}{5}, \dfrac{9}{5}, \dfrac{12}{5}, \ldots, \dfrac{36}{5}$ **21.** 4.5, 3, 1.5, … −36   **20.** _____

                                                                        21. _____

Name _____   Score _____

Find the sum of the indicated number of terms of the arithmetic sequence.

1. 3, 6, 9, …; $n = 20$

2. 1, 6, 11, …; $n = 30$

3. 20, 15, 10, …; $n = 40$

4. 2, 5, 8, …; $n = 25$

5. 4, 10, 16, …; $n = 20$

6. 20, $18\frac{1}{2}$, 17, …; $n = 10$

7. −4, 1, 6, …; $n = 16$

8. 50, 45, 40, …; $n = 10$

Simplify.

9. $\sum_{i=1}^{15} (3i + 2)$

10. $\sum_{i=1}^{10} (2i + 3)$

11. $\sum_{i=1}^{10} (3i - 2)$

12. $\sum_{n=1}^{16} \left(\frac{1}{2}n - 1\right)$

13. $\sum_{n=1}^{14} (2n + 5)$

14. $\sum_{n=1}^{14} (1 - 2n)$

15. $\sum_{i=1}^{15} (5 - 3i)$

16. $\sum_{n=1}^{10} (4 - n)$

1. _____
2. _____
3. _____
4. _____
5. _____
6. _____
7. _____
8. _____
9. _____
10. _____
11. _____
12. _____
13. _____
14. _____
15. _____
16. _____

Name

Score

Solve.

1. The distance that an object dropped from a helicopter will fall is 16 feet the first second, 48 feet the next second, 80 feet the third second, and so on in an arithmetic sequence. What is the total distance the object will fall in 5 seconds?

2. An exercise program calls for walking 15 minutes each day for a week. Each week thereafter, the amount of time spent walking increases by 5 minutes per day. In how many weeks will a person be walking 45 minutes each day?

1. _____

2. _____

3. A display of cans in a grocery store consists of 36 cans in the bottom row, 31 cans in the next row, and so on in an arithmetic sequence. The top row has 6 cans. Find the total number of cans in the display.

4. The salary schedule for an assistant manager of a computer store is $800 for the first month and includes a $40-per-month salary increase for the next ten months. Find the monthly salary for the tenth month. Find the total salary for the ten-month period.

3. _____

4. _____

5. An inventory of supplies for a calculator manufacturer indicated that 12,000 calculators were in stock on January 1. On February 1, and on the first day of each successive month, the manufacturer sent 800 calculators to retail stores. How many calculators were in stock after the shipment on September 1?

6. Some bricks are stacked so that there are 7 bricks in the top row and 3 more bricks in each successive row down to the ground. If there are 32 rows in all, how many bricks are there in the whole stack?

5. _____

6. _____

Name _____ Score _____

Find the indicated term of the geometric sequence.

1. $9, 3, 1, \ldots; a_7$

2. $3, 12, 48, \ldots; a_6$

3. $36, 24, 16, \ldots; a_8$

4. $1, 5, 25, \ldots; a_{10}$

5. $-1, \dfrac{1}{3}, -\dfrac{1}{9}, \ldots; a_6$

6. $1, 10, 100, \ldots; a_5$

7. $12, -4, \dfrac{4}{3}, \ldots; a_8$

8. $3, 6, 12, \ldots; a_6$

9. $\dfrac{1}{2}, \dfrac{1}{8}, \dfrac{1}{32}, \ldots; a_8$

1. _____

2. _____

3. _____

4. _____

5. _____

6. _____

7. _____

8. _____

9. _____

Find a_2 and a_3 for the geometric sequence.

10. $1, a_2, a_3, 27$

11. $2, a_2, a_3, 16$

12. $-3, a_2, a_3, 81$

13. $2, a_2, a_3, \dfrac{1}{4}$

14. $2, a_2, a_3, 250$

15. $3, a_2, a_3, -24$

16. $\dfrac{1}{16}, a_2, a_3, -\dfrac{1}{2}$

17. $4, a_2, a_3, -\dfrac{1}{2}$

18. $64, a_2, a_3, 1$

10. _____

11. _____

12. _____

13. _____

14. _____

15. _____

16. _____

17. _____

18. _____

Name _____   Score _____

Find the sum of the indicated number of terms of the geometric sequence.

1.   2, 4, 8, ...; $n = 6$ 2.   4, -2, 1, ...; $n = 5$

3. -4, 8, -16, ...; $n = 5$ 4.   27, 18, 12, ...; $n = 4$

5.   1, $\sqrt{3}$, 3, ...; $n = 8$ 6. $\dfrac{8}{27}$, $\dfrac{4}{9}$, $\dfrac{2}{3}$, ...; $n = 10$

Find the sum of the geometric series.

7. $\displaystyle\sum_{i=1}^{5} (5)^i$ 8. $\displaystyle\sum_{n=1}^{5} \left(\dfrac{1}{2}\right)^n$

9. $\displaystyle\sum_{i=1}^{4} \left(\dfrac{1}{4}\right)^i$ 10. $\displaystyle\sum_{n=1}^{4} \left(\dfrac{3}{4}\right)^n$

11. $\displaystyle\sum_{n=1}^{5} \left(\dfrac{1}{5}\right)^n$ 12. $\displaystyle\sum_{n=1}^{4} \left(\dfrac{2}{5}\right)^n$

1. _____

2. _____

3. _____

4. _____

5. _____

6. _____

7. _____

8. _____

9. _____

10. _____

11. _____

12. _____

Name

Score

Find the sum of the infinite geometric series.

1. $1 - \dfrac{1}{2} + \dfrac{1}{4} + \cdots$

2. $\dfrac{1}{2} + \dfrac{1}{4} + \dfrac{1}{8} + \cdots$

1. _____

2. _____

3. $4 - 2 + 1 + \cdots$

4. $1 + \dfrac{1}{4} + \dfrac{1}{16} + \cdots$

3. _____

4. _____

5. $2 + 1 + \dfrac{1}{2} + \cdots$

6. $\dfrac{1}{3} + \dfrac{1}{9} + \dfrac{1}{27} + \cdots$

5. _____

6. _____

7. $16 + 8 + 4 + \cdots$

8. $\dfrac{3}{10} + \dfrac{3}{100} + \dfrac{3}{1000} + \cdots$

7. _____

8. _____

Find an equivalent fraction for the repeating decimal.

9. $0.33\overline{3}$

10. $0.11\overline{1}$

11. $0.09\overline{09}$

9. _____

10. _____

11. _____

12. $0.15\overline{15}$

13. $0.63\overline{63}$

14. $0.27\overline{27}$

12. _____

13. _____

14. _____

15. $0.25\overline{25}$

16. $0.38\overline{8}$

17. $0.36\overline{36}$

15. _____

16. _____

17. _____

Name _____   Score _____

Solve.

1. A laboratory ore sample contains
    800 mg of a radioactive material with a
    half-life of 1 hour. Find the amount of
    radioactive material in the sample at the
    beginning of the fifth hour.

2. An ore sample contains 640 mg of a
    radioactive substance with a half-life
    of one day. Find the amount of
    radioactive material in the sample at the
    beginning of the fourth day.

1. _____

2. _____

3. A chain letter is sent to five people.
    Each of the five people in turn mails the
    letter to five other people, and the
    process is repeated. What is the total
    number of people who have received the
    letter after four mailings?

4. To test the "bounce" of a golf ball,
    the ball is dropped from a height of
    6 feet. The ball bounces 70% of its
    previous height with each bounce. How
    high does the ball bounce on the fifth
    bounce? Round to the nearest tenth.

3. _____

4. _____

5. The temperature of a hot water spa
    is 72°F. Each hour the temperature
    is 5% higher than during the previous
    hour. Find the temperature of the spa
    after 4 hours. Round to the nearest
    tenth.

6. A real estate broker estimates that
    a piece of land will increase in value
    at a rate of 10% each year. If the
    original value of the land is $20,000,
    what will be its value in 8 years?

5. _____

6. _____

Name _____ Score _____

Evaluate.

1. $2!$

2. $10!$

3. $6!$

4. $\dfrac{7!}{5!2!}$

5. $\dfrac{5!}{5!0!}$

6. $\dfrac{9!}{5!4!}$

7. $\dbinom{8}{2}$

8. $\dbinom{4}{3}$

9. $\dbinom{6}{5}$

10. $\dbinom{9}{8}$

11. $\dbinom{12}{1}$

12. $\dbinom{8}{6}$

Write in expanded form.

13. $(x+y)^3$

14. $(x-y)^4$

15. $(x-1)^4$

16. $(2x-3)^3$

Find the first three terms of the expansion.

17. $(a+b)^6$

18. $(a-b)^7$

19. $(2x-3y)^6$

20. $(3x+y)^5$

Find the indicated term in the expansion.

21. $(x+3)^5$; 3rd term

22. $(y-1)^8$; 2nd term

23. $\left(x-\dfrac{1}{2}\right)^4$; 1st term

24. $(x-4)^6$; 6th term

Score

1. _____
2. _____
3. _____
4. _____
5. _____
6. _____
7. _____
8. _____
9. _____
10. _____
11. _____
12. _____
13. _____
14. _____
15. _____
16. _____
17. _____
18. _____
19. _____
20. _____
21. _____
22. _____
23. _____
24. _____

CHAPTER 1

Objective 1.1A

1. 3   **2.** 7   **3.** −5   **4.** −17   **5.** 87   **6.** −72   **7.** −66
8. −52   **9.** 120   **10.** 36   **11.** −3, 4   **12.** −8, −2   **13.** 60, −13, −28   **14.** 48, 5, 9

Objective 1.1B

1. {−2, −1, 0, 1}   **2.** {1, 3, 5, 7}   **3.** {5, 10, 15, 20, 25, 30, 35}   **4.** {−18, −12, −6}
5. $\{x \mid x \geq -6\}$ **6.** $\{x \mid -11 < x < -2\}$ **7.** $\{x \mid x < 10\}$ **8.** $\{x \mid -1 \leq x \leq 1\}$

9.   **10.**
11.   **12.**

13. $[-3, \infty)$ **14.** $(-4, -2]$ **15.** $\{x \mid x < -1\}$ **16.** $\{x \mid 4 \leq x < 7\}$
17.   **18.**
19.   **20.**

Objective 1.1C

1. $A \cup B = \{-2, -1, 0, 1, 2\}$ **2.** $A \cup B = \{-3, -1, 0, 1, 3\}$ **3.** $A \cup B = \{-2, -1, 0, 2\}$
4. $A \cap B = \{0\}$ **5.** $A \cap B = \{5, 10\}$ **6.** $A \cap B = \{1, 4, 9\}$

7.   **8.**
9.   **10.**
11.   **12.**
13.   **14.**
15.   **16.**
17.   **18.**

Objective 1.2A

1. −26   **2.** −27   **3.** −9   **4.** −240   **5.** −60   **6.** 0   **7.** −6
8. −5   **9.** 6   **10.** 120   **11.** 390   **12.** −16,200   **13.** 37,840   **14.** 80
15. 84   **16.** 4   **17.** 20   **18.** 8   **19.** 4   **20.** 7   **21.** 13
22. −26   **23.** −1   **24.** 22   **25.** 1   **26.** −3   **27.** −18

Objective 1.2□

1. 16   **2.** −125   **3.** −64   **4.** 81   **5.** 49   **6.** −243   **7.** −81
8. −125   **9.** 256   **10.** 36   **11.** 36   **12.** 256   **13.** 108   **14.** −72
15. 576   **16.** −72   **17.** −72   **18.** 225   **19.** −200   **20.** −108   **21.** 144
22. −400   **23.** 8000   **24.** −144   **25.** −160   **26.** 160   **27.** 288

Objective 1.3B

1. $1\frac{5}{24}$ **2.** $-\frac{13}{48}$ **3.** $-\frac{23}{15}$ **4.** $\frac{1}{4}$ **5.** $\frac{13}{24}$ **6.** $\frac{1}{15}$ **7.** $\frac{11}{32}$

8. $-\frac{5}{12}$ **9.** $-\frac{5}{16}$ **10.** $-\frac{3}{4}$ **11.** 1   **12.** $-\frac{3}{4}$ **13.** $-\frac{1}{20}$ **14.** $\frac{1}{6}$

15. $-\frac{2}{3}$ **16.** −25.63   **17.** −9.56   **18.** −7.976   **19.** 138.039   **20.** −5.504   **21.** 10.01

22. 0.946   **23.** −4.83   **24.** 4.35   **25.** −5   **26.** −7.566   **27.** −26.482

Objective 1.□D

1. −5   **2.** −50   **3.** $17\frac{3}{5}$ **4.** 12   **5.** $9\frac{6}{7}$ **6.** $\frac{5}{84}$ **7.** −92

8. 26   **9.** 45   **10.** 17   **11.** 18   **12.** 56   **13.** $-\frac{1}{12}$ **14.** $\frac{55}{48}$

15. $-\frac{23}{24}$ **16.** $\frac{11}{6}$ **17.** 4.892   **18.** 2.2

Objective 1.4A

1. 6   **2.** 8   **3.** 1   **4.** $3 \cdot 7$ **5.** $-x$ **6.** 0   **7.** 0
8. wz **9.** 20   **10.** 1   **11.** 0   **12.** 3   **13.** The Inverse Property of Addition
14. The Distributive Property   **15.** The Multiplication Property of Zero   **16.** The Multiplication Property of One
17. The Associative Property of Addition   **18.** The Division Properties of Zero   **19.** The Division Properties of Zero
20. The Addition Property of Zero   **21.** The Associative Property of Multiplication
22. The Inverse Property of Multiplication   **23.** The Commutative Property of Addition
24. The Commutative Property of Multiplication

Objective 1.4B

1. −10   **2.** 25   **3.** −2   **4.** −1   **5.** −16   **6.** 45   **7.** $\frac{1}{8}$

8. 12   **9.** 5   **10.** −3   **11.** $\frac{3}{2}$ **12.** 1   **13.** 4   **14.** −9

15. −22   **16.** 3   **17.** −4   **18.** 12   **19.** 10   **20.** −9   **21.** 2
22. 2   **23.** −5   **24.** 5   **25.** 13   **26.** −18   **27.** −8

Objective 1.4C

1. $11x$ **2.** $14x$ **3.** $-5x$ **4.** $7x$ **5.** $8a + 5b$ **6.** $-9a - 9b$ **7.** y
8. x **9.** $x + y$ **10.** $x - y$ **11.** $-4a - 12$ **12.** $-11x + 18$ **13.** $2x + 24y$ **14.** $105a - 80$
15. $-8x + 24y$ **16.** $-9x - 12y$ **17.** $-6x - 24y$ **18.** $-24x - 4y$ **19.** $12x - 14y$ **20.** $28a - 37b$ **21.** $-13a - 4b$
22. $51x - 65y$ **23.** $x - 18y - 30$ **24.** $5 - 5x + y$ **25.** $5x - 8$ **26.** $3x + 9$ **27.** $3x + 6$

Objective 1.5A

1. $n + 2n$; $3n$ **2.** $n + (6 - n)$; 6   **3.** $n - \frac{1}{3}n$; $\frac{2}{3}n$ **4.** $\frac{1}{3}(9n + 21)$; $3n + 7$ **5.** $9\left(\frac{n}{3}\right)$; $3n$ **6.** $\frac{1}{2}n + \frac{3}{4}n$; $\frac{5}{4}n$

7. $12(2n)$; $24n$ **8.** $n - \frac{3}{4}n$; $\frac{1}{4}n$ **9.** $n - (n - 4)$; 4   **10.** $4n + 2(n - 4)$; $6n - 8$ **11.** $2[n + (n + 2)]$; $4n + 4$

12. $20 - \frac{1}{3}(n + 9)$; $17 - \frac{1}{3}n$ **13.** $(n + 2) + (n + 4)$; $2n + 6$ **14.** $\frac{1}{3}(n + n + 2 + n + 4)$; $n + 2$

15. $[n + (n + 2)] - 1$; $2n + 1$ **16.** $2(n + 11) + 7$; $2n + 28$ **17.** $9n - (4n + 6)$; $5n - 6$

18. $4 + (n + 5)9$; $9n + 49$ **19.** $10(6 + n) - n$; $60 + 9n$ **20.** $11 - (n - 2)10$; $31 - 10n$

Objective 1.5B

1. Length of shorter piece: L; $10 - L$ **2.** Measure of angle A: x; Measure of angle B: $5x$; Measure of angle C: $50x$
3. Length of shorter piece: L; $3 - L$ **4.** $20{,}000 - d$ **5.** Amount of peanuts: p; $p + 2$
6. Length of tunnel: L; $L + 1883$ **7.** Number of singles: S; $4S$ **8.** Yellow paint: Y; $2Y$

CHAPTER 2

Objective 2.1A

1. 9	**2.** 10	**3.** −5	**4.** 6	**5.** −4	**6.** −9	**7.** 5
8. $\frac{1}{2}$	**9.** $-\frac{5}{2}$	**10.** $-\frac{7}{3}$	**11.** 21	**12.** $\frac{9}{16}$	**13.** $\frac{2}{15}$	**14.** $\frac{1}{10}$
15. 12	**16.** 10	**17.** −12	**18.** −81	**19.** $-\frac{18}{25}$	**20.** $-\frac{10}{21}$	**21.** $-\frac{5}{18}$
22. $\frac{3}{4}$	**23.** −49	**24.** −20	**25.** 3.25	**26.** 15.38	**27.** −3.84	

Objective 2.1B

1. 3	**2.** −4	**3.** −1	**4.** 0	**5.** 3	**6.** −2	**7.** 5
8. −3	**9.** −2	**10.** 9	**11.** −2	**12.** $\frac{5}{3}$	**13.** 2	**14.** 1
15. 8	**16.** 12	**17.** $\frac{10}{3}$	**18.** 1	**19.** −1	**20.** $-\frac{2}{3}$	**21.** −4
22. 8	**23.** −2	**24.** 3	**25.** 8	**26.** 2	**27.** 14	

Objective 2.1C

1. −6	**2.** 2	**3.** 2	**4.** 2	**5.** 3	**6.** 2	**7.** 1
8. 9	**9.** −5	**10.** 13	**11.** −6	**12.** 4	**13.** 36	**14.** 11
15. −7	**16.** 1	**17.** −14	**18.** 1	**19.** $\frac{3}{11}$	**20.** $\frac{1}{12}$	

Objective 2.1D

1. $n = \dfrac{a_n - a_1}{d} + 1$ **2.** $b = \dfrac{2A}{h}$ **3.** $a = -\dfrac{b}{2a}$ **4.** $t = \dfrac{A-P}{Pr}$ **5.** $n = \dfrac{PV}{RT}$ **6.** $h = \dfrac{2A}{b}$

7. $G = \dfrac{Fr^2}{m_1 m_2}$ **8.** $V_2 = \dfrac{P_1 V_1 T_2}{P_2 T_1}$ **9.** $R = \dfrac{E - Ir}{I}$ **10.** $b_1 = \dfrac{2A - hb_2}{h}$ **11.** $r = \dfrac{D}{t}$ **12.** $C = \dfrac{100M}{I}$

13. $T = \dfrac{E + alt}{al}$ **14.** $a = \dfrac{bt}{b - t}$ **15.** $v = \dfrac{2at - 2d}{a}$ **16.** $n = \dfrac{R - C}{P}$ **17.** $w = \dfrac{S - 2hL}{2h + 2L}$ **18.** $g = \dfrac{2h - 2vt}{t^2}$

Objective 2.2A

1. 18 lb of $7 and 27 lb of $5   **2.** $2.25/lb   **3.** 320   **4.** 300   **5.** 200 bushels of soy, 800 bushels of wheat
6. 112 L   **7.** 20 oz   **8.** 22.5 oz of pure gold; 37.5 oz of the alloy   **9.** $6.60/oz   **10.** $4.35/lb
11. 90 lb   **12.** 30 gal

Objective 2.2B

1. 46%   **2.** 28%   **3.** 24 ml   **4.** 50 oz   **5.** 20 lb of 24%; 60 lb of 16%
6. 11.25 L of 70%; 18.75 L of 30%   **7.** 75 g   **8.** 640 lb   **9.** 45 L of 70%; 75 L of 30%
10. 450 lb of 20%; 150 lb of 80%   **11.** 30 ml   **12.** 75 ml

Objective 2.2C

1. express bus: 65 mph; local bus: 50 mph   **2.** 1st plane: 260 mph; 2nd plane 290 mph   **3.** 140 mi   **4.** 78 mi
5. 350 mi   **6.** 54 mi   **7.** 1st car: 56 mph; 2nd car: 50 mph   **8.** 1st plane: 320 mph; 2nd plane: 280 mph
9. 1st plane: 180 mph; 2nd plane: 140 mph   **10.** 40 mi

Objective 2.3A

1. $\{x \mid x < 6\}$ **2.** $\{x \mid x \geq -3\}$ **3.** $\{x \mid x \leq 5\}$ **4.** $\{x \mid x > 5\}$ **5.** $\{x \mid x < -3\}$ **6.** $\{x \mid x \geq 4\}$ **7.** $\{x \mid x > 2\}$
8. $\{x \mid x \geq -2\}$ **9.** $\{x \mid x > 3\}$ **10.** $\{x \mid x < 6\}$ **11.** $\{x \mid x \leq 3\}$ **12.** $\{x \mid x \leq -2\}$ **13.** $\{x \mid x > -3\}$ **14.** $\{x \mid x < -5\}$
15. $\{x \mid x > -9\}$ **16.** $\{x \mid x < -2\}$ **17.** $\{x \mid x \geq 4\}$ **18.** $\{x \mid x < 2\}$ **19.** $\{x \mid x \leq -4\}$ **20.** $\{x \mid x \geq -3\}$ **21.** $\{x \mid x < 1\}$
22. $\{x \mid x > -8\}$ **23.** $\{x \mid x < -4\}$ **24.** $\{x \mid x \geq -5\}$

Objective 2.3B

1. $\{x \mid -3 < x < 5\}$ **2.** $\{x \mid -5 < x \leq 8\}$ **3.** $\{x \mid x \geq 6 \text{ or } x < 3\}$ **4.** $\{x \mid x < 4 \text{ or } x > 9\}$
5. $\{x \mid -3 < x < 6\}$ **6.** $\{x \mid -3 < x < 3\}$ **7.** $\{x \mid x \geq 4\}$ **8.** $\{x \mid x < -5\}$ **9.** $\varnothing$ **10.** $\{x \mid -2 < x < 2\}$
11. $\{x \mid x < 1 \text{ or } x > 3\}$ **12.** $\{x \mid x < -3 \text{ or } x > 3\}$ **13.** $(-3, 1)$ **14.** $(-2, 4)$ **15.** $(4, 5)$ **16.** $(-3, -1)$
17. $(-\infty, -2] \cup (3, \infty)$ **18.** $(-\infty, -1) \cup (3, \infty)$ **19.** $(4, \infty)$ **20.** $(-\infty, 2)$ **21.** $[-4, 3]$ **22.** $[3, 5]$
23. $[-2, 1)$ **24.** $\varnothing$

Objective 2.3C

1. 3 ft   **2.** −19   **3.** less than 2501 miles   **4.** $-5° < C < 20°$ **5.** $32,000 or more
6. less than 30 checks   **7.** $53 \leq N \leq 98$ **8.** 25, 27, 29 or 27, 29, 31 or 29, 31, 33

Objective 2.4A

1. 6 and −6   **2.** 3 and −3   **3.** 5 and −5   **4.** 11 and −11   **5.** 7 and −7   **6.** 2 and −2   **7.** 9 and −9
8. 1 and −1   **9.** No solution   **10.** No solution   **11.** 2   **12.** −1 and 7   **13.** 6 and −2   **14.** 6
15. $\dfrac{2}{3}$ **16.** 4 and −1   **17.** $-\dfrac{5}{3}$ and 3   **18.** −3 and −4   **19.** 1 and $-\dfrac{3}{5}$ **20.** $\dfrac{3}{2}$ and 0   **21.** $\dfrac{2}{3}$ and −2
22. $\dfrac{3}{2}$ and −3   **23.** No solution   **24.** $-\dfrac{1}{3}$ **25.** 2 and −1   **26.** $\dfrac{5}{3}$ and $-\dfrac{1}{3}$ **27.** $\dfrac{5}{2}$

Objective 2.4B

1. $\{x \mid x > 2 \text{ or } x < -2\}$ **2.** $\{x \mid -4 < x < 4\}$ **3.** $\{x \mid x \leq 2 \text{ or } x \geq 4\}$ **4.** $\{x \mid x \geq 7 \text{ or } x \leq -3\}$
5. $\{x \mid -3 < x < 4\}$ **6.** $\{x \mid x \geq -4 \text{ or } x \leq -10\}$ **7.** $\{x \mid 1 < x < 5\}$ **8.** $\{x \mid -3 \leq x \leq 11\}$
9. $\left\{x \mid x > 2 \text{ or } x < -\dfrac{16}{3}\right\}$ **10.** $\left\{x \mid x > 3 \text{ or } x < -\dfrac{13}{5}\right\}$ **11.** $\varnothing$ **12.** $\left\{x \mid 1 \leq x \leq \dfrac{5}{3}\right\}$ **13.** $\left\{x \mid x < -2 \text{ or } x > \dfrac{7}{2}\right\}$
14. $\left\{x \mid -\dfrac{8}{3} < x < 4\right\}$ **15.** $\{x \mid x = 9\}$ **16.** $\left\{x \mid -\dfrac{9}{5} < x < 3\right\}$ **17.** $\{x \mid x < -2 \text{ or } x > 5\}$ **18.** $\left\{x \mid -\dfrac{1}{2} < x < 1\right\}$

Objective 2.4C

1. 3.94 cc; 4.06 cc   **2.** 85 volts; 125 volts   **3.** 97.5 volts; 132.5 volts   **4.** 2.47 cc; 2.53 cc
5. 27,440 ohms; 28,560 ohms   **6.** 14,720 ohms; 17,280 ohms   **7.** 68.4 ohms; 75.6 ohms   **8.** 5.75 volts; 6.75 volts

CHAPTER 3

Objective 3.1A

1. 5.39   **2.** 5   **3.** 6.08   **4.** 4.47   **5.** (–1, 0)   **6.** $\left(\frac{5}{2}, \frac{3}{2}\right)$ **7.** (0, 3)

8. $\left(\frac{1}{2}, \frac{1}{2}\right)$ **9.** (1, 1)   **10.** $\left(\frac{3}{2}, -\frac{3}{2}\right)$

Objective 3.1B

1.   **2.**   **3.**   **4.**

5. *A* is (2, 0), *B* is (–3, 1), *C* is (1, 3), *D* is (–2, -3)   **6.** *A* is (–4, 1), *B* is (–3, –2), *C* is (3, 2), *D* is (1, 4)

7.   **8.**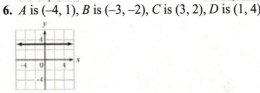

Objective 3.2A

1. Yes   **2.** No   **3.** –1   **4.** 5   **5.** 1   **6.** –1   **7.** 1

8. –1   **9.** 5   **10.** 1   **11.** $t^2 + t - 1$ **12.** Domain: {1, 2, 4, 5}; Range: {1, 3, 7}

13. Domain: {1, 3, 5, 6}; Range: {1, 2, 4}   **14.** none   **15.** –5   **16.** 3   **17.** Domain: {–1, 1, 3}; Range: {–1, 7, 15}

18. Domain: {–1, 0, 3}; Range: $\left\{\frac{3}{5}, \frac{3}{4}, 1\right\}$

Objective 3.3A

1. 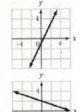   **2.** 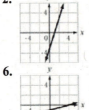   **3.**   **4.**

5.   **6.**   **7.**   **8.**

Objective 3.3B

1.

2.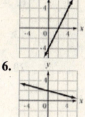

3.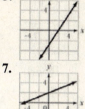

4.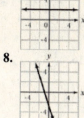

5.

6.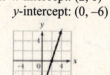

7.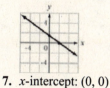

8.

Objective 3.3C

1. x-intercept: (5, 0)
   y-intercept: (0, −2)

2. x-intercept: (2, 0)
   y-intercept: (0, −6)

   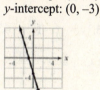

3. x-intercept: (2, 0)
   y-intercept: $\left(0, \frac{4}{3}\right)$

4. x-intercept: (1, 0)
   y-intercept: (0, 2)

5. x-intercept: (−6, 0)
   y-intercept: (0, −2)

6. x-intercept: (−1, 0)
   y-intercept: (0, −3)

7. x-intercept: (0, 0)
   y-intercept: (0, 0)

8. x-intercept: $\left(\frac{5}{2}, 0\right)$
   y-intercept: (0, −5)

Objective 3.3D

1. Loren earns $160 for tutoring 16 hours.

   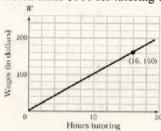

2. The roller coaster travels 594 ft in 6 s.

   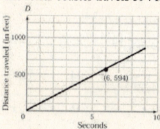

3. The cost of receiving 36 messages is $16.

   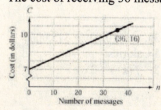

4. The cost of manufacturing 60 snow boards is $9000.

   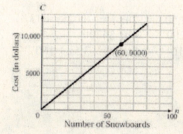

Objective 3.4A

1. $\frac{1}{2}$ 2. $-\frac{1}{2}$ 3. -3 4. 1 5. 1 6. $-\frac{1}{2}$ 7. -1

8. $-\frac{1}{3}$ 9. $-\frac{5}{3}$ 10. $\frac{5}{2}$ 11. 1 12. 2 13. -1 14. -5

15. -2 16. $\frac{1}{3}$ 17. -3 18. -1 19. 0 20. $\frac{1}{4}$ 21. 1

Objective 3.4C

1.   2.   3.   4.

5.   6.   7.   8.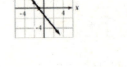

Objective 3.5A

1. $y = 2x + 4$ 2. $y = 4x + 8$ 3. $y = \frac{1}{3}x + \frac{5}{3}$ 4. $y = -\frac{1}{3}x - 1$ 5. $y = -x - 1$ 6. $y = \frac{1}{2}x + \frac{7}{2}$ 7. $y = \frac{2}{3}x + \frac{5}{3}$

8. $y = -\frac{1}{3}x + 3$ 9. $y = -2x + 3$ 10. $y = -\frac{2}{5}x + \frac{13}{5}$ 11. $y = \frac{3}{4}x - \frac{11}{4}$ 12. $y = \frac{1}{3}x - \frac{8}{3}$

13. $y = -\frac{1}{2}x - \frac{3}{2}$ 14. $y = -\frac{1}{4}x$ 15. $y = 2x - 8$ 16. $y = -2x + 2$ 17. $y = -\frac{1}{3}x + \frac{14}{3}$

18. $y = -\frac{3}{5}x + \frac{17}{5}$

Objective 3.5B

1. $y = -x + 5$ 2. $y = -x + 4$ 3. $y = \frac{2}{3}x + \frac{4}{3}$ 4. $y = \frac{1}{2}x - \frac{1}{2}$ 5. $y = \frac{1}{3}x + \frac{8}{3}$ 6. $y = x$ 7. $y = \frac{3}{2}x$

8. $y = \frac{4}{3}x + \frac{11}{3}$ 9. $y = 2x - 2$ 10. $y = 3x + 3$ 11. $y = \frac{1}{2}x$ 12. $y = -3x$ 13. $y = -\frac{2}{3}x + \frac{2}{3}$

14. $y = -\frac{4}{3}x - \frac{4}{3}$ 15. $y = -\frac{6}{5}x + \frac{8}{5}$ 16. $y = -\frac{6}{7}x - \frac{11}{7}$ 17. $y = -x + 2$ 18. $y = -2$

Objective 3.5C

1a. $y = 95x + 35{,}000$ **1b.** \$177,500   **2a.** $y = -0.025x + 13$ **2b.** 8.5 gal   **3a.** $y = -0.10x + 315{,}000$
3b. 82,500 cars   **4a.** $y = 0.05x + 1200$ **4b.** \$4700   **5a.** $y = 22.5x$ **5b.** 112.5 Calories
6a. $y = 0.69x + 7.95$ **6b.** \$14.16   **7a.** $y = -3.5x + 100$ **7b.** 82.5°C   **8.** $f(x) = -2x - 7$

Objective 3.6A

1. Yes **2.** No **3.** No **4.** Yes **5.** Yes **6.** Yes **7.** Yes

8. No **9.** Yes **10.** $y = -\frac{1}{3}x - \frac{17}{3}$ **11.** $y = \frac{1}{4}x - \frac{7}{4}$ **12.** $y = \frac{5}{3}x - \frac{11}{3}$

13. $y = -\frac{1}{2}x - \frac{1}{2}$ **14.** $y = -\frac{4}{3}x + \frac{8}{3}$ **15.** $y = \frac{2}{3}x - \frac{2}{3}$ **16.** $y = -3x + 2$

Objective 3.7A

1.   **2.**   **3.**   **4.**

5.   **6.**   **7.**   **8.**

CHAPTER 4

Objective 4.1A

1. The solution is (–3, 2). **2.** The solution is (2, –1). **3.** The solution is (–2, –3). **4.** The solution is (3, 1).

5. The solution is (2, –2). **6.** The solution is (3, 2). **7.** The solution is (2, 1). **8.** The solution is (–1, 3).

Objective 4.1B

1. (3, 2) **2.** (4, –1) **3.** (3, 1) **4.** (–2, 0) **5.** (5, 3) **6.** (2, 1) **7.** (1, –2)

8. (–2, 3) **9.** (2, 1) **10.** (–2, –1) **11.** (–1, 3) **12.** (2, –3) **13.** (–1, 4) **14.** (–2, –1)

15. (4, 1) **16.** $\left(1, -\frac{1}{2}\right)$ **17.** (2, –2) **18.** (1, 1) **19.** (–5, -2) **20.** (–4, –2) **21.** (1, 3)

22. (1, 1) **23.** (2, 5) **24.** (–2, 7)

Objective 4.1C

1. $6400 @ 4.5%; $3600 @ 6% **2.** $8000 @ 10%; $6000 @ 14.5% **3.** $2500 **4.** $6000

5. $6500 @ 11%; $5500 @13% **6.** $14,250 @ 15%; $23,750 @9% **7.** $14,000 **8.** $10,000

9. $6000 **10.** $14,500

Objective 4.2A

1. (4, 1)   **2.** (3, 2)   **3.** (4, –3)   **4.** (1, –1)   **5.** (–1, 3)   **6.** (–2, 2)   **7.** (5, 1)

8. (3, 5)   **9.** (4, –2)   **10.** (2, 5)   **11.** (2, –1)   **12.** (3, 2)   **13.** $\left(\frac{1}{2}, 1\right)$ **14.** (–2, 1)

15. (1, –3)   **16.** (5, –1)   **17.** (–1, –2)   **18.** (2, –3)   **19.** (–1, 2)   **20.** (4, –2)   **21.** (–2, 1)

22. (4, –1)   **23.** $\left(-\frac{1}{7}, -\frac{25}{7}\right)$ **24.** (20, 15)

Objective 4.2B

1. (–2, 1, 3)   **2.** (4, –2, 1)   **3.** (–5, 3, –2)   **4.** (–3, 4, 1)   **5.** (3, –4, 2)   **6.** (–1, 3, 4)   **7.** (1, –2, –1)

8. (–2, 3, –1)   **9.** $\left(\frac{162}{11}, \frac{65}{11}, -6\right)$ **10.** (1, –3, –4)

Objective 4.3A

1. 10   **2.** –7   **3.** 12   **4.** –11   **5.** –19   **6.** –13   **7.** 34
8. 1   **9.** 2   **10.** –50   **11.** –21   **12.** –32   **13.** –26   **14.** 32
15. 16   **16.** 7   **17.** –51   **18.** 2

Objective 4.3B

1. (1, 8)   **2.** (0, –1)   **3.** (4, 1)   **4.** (3, –1)   **5.** (3, 7)   **6.** (1, –1)   **7.** (2, 2)

8. (–3, –5)   **9.** $\left(-\frac{1}{3}, 2\right)$ **10.** (0, 2)   **11.** (–1, 1, 4)   **12.** (1, 2, 5)   **13.** (2, –1, 1)   **14.** (–1, 1, 3)

15. (–1, 2, 1)   **16.** (3, 1, 0)

Objective 4.4A

1. plane: 170 mph; wind: 35 mph   **2.** plane: 520 mph; wind: 30 mph   **3.** cruiser: 15 mph; current: 5 mph
4. motorboat: 15 mph; current: 3 mph   **5.** plane: 180 mph; wind: 20 mph   **6.** jet: 225 mph; wind: 25 mph
7. boat: 24 km/h; current: 4 km/h   **8.** team: 12 mph; current: 3 mph   **9.** plane: 140 mph; wind: 20 mph
10. plane: 315 mph; wind: 45 mph

Objective 4.4B

1. pine: $0.20/ft; redwood: $0.30/ft   **2.** cinnamon: $2/lb; spice: $3.50/lb   **3.** 10¢/unit   **4.** $12/yd
5. 40 quarters   **6.** 18 dimes   **7.** $3/lb   **8.** tin: $3.50/lb; zinc: $9/lb
9. orange trees: $8; grapefruit trees: $6   **10.** bulbs: 50¢; lights: $2

Objective 4.5A

1.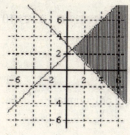

2.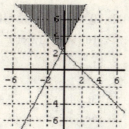

3.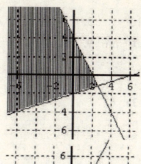

4.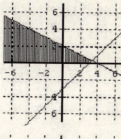

5.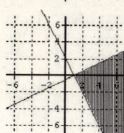

6.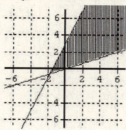

7.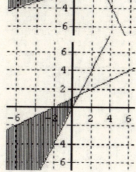

8.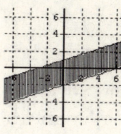

CHAPTER 5

Objective 5.1A

1. $a^4 b^4$
2. $12 x^4 y^5$
3. $x^2 y^6$
4. $x^9 y^6$
5. $81 a^8 b^4$
6. $-64 x^6 y^9$
7. $64 a^{12} b^6$
8. $x^5 y^4$
9. $x^7 y^3$
10. $-200 a^8 b^5 c^6$
11. $-100 a^8 b^9$
12. x^{6n}
13. x^{2n-2}
14. x^{7x+2}
15. x^{5n-4}
16. a^{3n^2}
17. a^{3n^2-6n}
18. x^{6n-2}
19. x^{16n+4}
20. a^{3n^2-n}
21. $60 x^8 y^3 z^4$
22. $-160 x^7 y^6 z^6$
23. $-18 x^4 y^4 z^4$
24. $12 a^5 b^3 c^4$

Objective 5.1B

1. $\dfrac{1}{32}$
2. $\dfrac{a^2 b}{3}$
3. $\dfrac{x}{y^3}$
4. $-\dfrac{1}{9}$
5. $-\dfrac{1}{8}$
6. a
7. $\dfrac{1}{x^5}$
8. x
9. $\dfrac{a^5}{16}$
10. $\dfrac{1}{x^4 y^3}$
11. $x^5 y^3$
12. $\dfrac{1}{a^5 b}$
13. $3a^3$
14. $-\dfrac{x^2}{4y^2}$
15. $\dfrac{bc^2}{a^2}$
16. $\dfrac{b^2}{a^4 c^4}$
17. y^2
18. x
19. $-\dfrac{1}{2ab^8}$
20. $-\dfrac{8a^3}{27b^3}$
21. $\dfrac{81 x^4 z^8}{256 y^4}$
22. $\dfrac{y}{24 x^4}$
23. $-\dfrac{y}{4x}$
24. $\dfrac{1}{a^n}$
25. $-\dfrac{1}{a^{3n}}$
26. x^{n-1}
27. $a^{n-2} b^{n+1}$

Objective 5.1C

1. 3.56×10^{-5}
2. 3.2×10^5
3. 3.7×10^{-7}
4. 5.1×10^{-9}
5. 6×10^{10}
6. 4.6×10^7
7. 0.00000214
8. 0.0000000053
9. $6,700,000,000$
10. $5,230,000$
11. 0.0045
12. $52,400,000$
13. 0.028
14. 0.00000001638
15. $350,000$
16. 0.0000002144
17. $31,490$
18. 0.2232
19. 0.000000004
20. $70,000,000$
21. 0.000000000018
22. $3,000,000,000$
23. 600
24. 0.55

Objective 5.1D

1. 1.404×10^{10} km   **2.** 7.5×10^6 **3.** 5.1×10^9 **4.** 1.3×10^6 s   **5.** 9.6×10^8 revolutions   **6.** 3.52×10^{-7} s
7. 3.52×10^{-6} s   **8.** 4.2×10^6 mi   **9.** 6.428×10^{10} mi   **10.** 4.82112×10^{11} mi

Objective 5.2A

1. -129 **2.** 22   **3a.** 6   **3b.** 7   **3c.** 4   **4.** not a polynomial   **5a.** 1
5b. $-\sqrt{7}$ **5c.** 2   **6a.** π **6b.** π **6c.** 0
7.   **8.**   **9.**

Objective 5.2B

1. $x^2 - 4x - 4$ **2.** $3x^3 + 2x^2 - 16x - 4$ **3.** $4x^2 - x - 4$ **4.** $-2x^2 + 3xy + y^2$ **5.** $x^{2n} + 5x^n + 3$
6. $5x^{2n} - 2x^n + 6n + 1$ **7.** $4y^3 + y^2 - 15y + 1$ **8.** $3y^2 - 8y - 11$ **9.** $6a^2 - a + 2$ **10.** $4x^4 + 2x^3 - 5x^2 + 3x$
11. $2x^4 + 4x^3 - 7x - 5$ **12.** $-2a^{2n} - a^n - 11$ **13.** $-2x^{2n} - x^n - 11$ **14.** $-3x^3 + 9x^2 + x - 11$
15. $3a^3 - 8a^2b + 8ab^2 + 11b^3$

Objective 5.3A

1. $3x^2 - 6x$ **2.** $4a^2 + 12a$ **3.** $8x^2y - 10xy^2$ **4.** $-10a^2b + 15ab^2$ **5.** $x^{3n} - x^n$ **6.** $x^{3n} - 4x^{2n}$
7. $x^{2n} - x^n y^n$ **8.** $4x - 3x^2$ **9.** $-3a^2 + 10a$ **10.** $-6a^4 + 9a^3 - 12a^2$ **11.** $10b^4 - 50b^3 - 20b$
12. $8b^5 - 10b^3 + 14b$ **13.** $-9x^4 + 6x^3 + 18x^2$ **14.** $-4y^5 - 3y^4 + 4y^3$ **15.** $-15b^5 + 12b^4 + 6b^2$
16. $-12x^2 + 8x^3 - 8x^4 - 12x^5$ **17.** $-6y + 9y^2 + 12y^3 - 9y^4$ **18.** $a^{2n-1} - 2a^n + 3a^{n-1}$
19. $a^{2n+2} + 5a^{n+4} - 2a^{n+3}$ **20.** $6y^2 - 21y$ **21.** $-2a + 6a^2 - a^3 - a^4$

Objective 5.3B

1. $x^2 + 5x - 6$ **2.** $y^2 + 9y + 14$ **3.** $2y^2 - 5y - 3$ **4.** $6x^2 - 13x + 5$ **5.** $12a^2 + 23ab + 10b^2$ **6.** $10x^2 - 19xy - 15y^2$
7. $20x^2 - 51xy + 28y^2$ **8.** $x^2y^2 + 2xy - 24$ **9.** $x^4 - 8x^2 + 15$ **10.** $6x^4 - 7x^2y + 2y^2$ **11.** $x^4 + 2x^2y^2 - 3y^4$
12. $x^{2n} - x^n - 12$ **13.** $x^{2n} - 7x^n + 12$ **14.** $6a^{2n} + a^n - 2$ **15.** $8x^{2n} - 2x^n - 3$
16. $6a^{2n} + a^n b^n - 2b^{2n}$ **17.** $4a^{2n} - 1$ **18.** $x^3 - 3x^2 + 7x - 5$ **19.** $x^4 + 3x^3 - 2x^2 - 4x + 6$
20. $a^4 - a^3 - 2a^2 + 5a + 5$ **21.** $x^4 - 4x^3 + x^2 - 7x + 12$ **22.** $a^3 - 5a^2 + 2a + 8$
23. $y^5 - 4y^4 - 2y^3 + 6y^2 + 4$ **24.** $x^5 + 3x^3 - 3x^2 + 2x - 3$

Objective 5.3C

1. $a^2 - 9$ 2. $b^2 - 64$ 3. $16x^2 - 25y^2$ 4. $25x^2 - 9y^2$ 5. $x^2 y^2 - 16$ 6. $16a^2 - 9c^2$ 7. $4a^2 - b^2$

8. $x^4 - 16$ 9. $x^4 - 49$ 10. $x^{2n} - 1$ 11. $a^{2n} - b^2$ 12. $x^2 - 16y^2$ 13. $x^2 - 12x + 36$

14. $16x^2 - 8xy + y^2$ 15. $x^4 - 18x^2 + 81$ 16. $x^2 y^2 + 6xy + 9$ 17. $x^2 y^2 - 14xy + 49$

18. $x^4 - 4x^2 y^2 + 4y^4$ 19. $x^4 - 2x^2 + 1$ 20. $x^4 - 2x^2 y^2 + y^4$ 21. $9x^4 + 6x^2 y^2 + y^4$ 22. $x^4 - 8x^2 + 16$

23. $x^4 - 14x^2 + 49$ 24. $4x^4 + 20x^2 y^2 + 25y^4$ 25. $x^{2n} + 2x^n + 1$ 26. $a^{2n} + 2a^n b^n + b^{2n}$

27. $9x^{2n} + 24x^n y^n + 16y^{2n}$

Objective 5.3D

1. $4x^2 + 9x - 20$ ft^2 2. $x^2 - x - 20$ ft^2 3. $9x^2 + 12x + 4$ ft^2 4. $x^3 - 9x^2 + 27x - 27$ cm^3

5. $x^2 + 5x$ m^2 6. $x^2 + 18x + 36$ ft^2 7. $3x^3$ in.^{3} 8. $4x^3 + 22x^2 + 10x$ cm^3

9. $12.5x^2 - 12.56x + 3.14$ in.^{2} 10. $3.14x^2 + 12.56x + 12.56$ in.^{2}

Objective 5.4A

1. $1 + 3x$ 2. $3x - 1$ 3. $2x^2 - x + 3$ 4. $x^2 y^2 - 9xy + 3$ 5. $a^2 + 2a - 3$ 6. $x - 3$

7. $2x + 3$ 8. $-5x^2 + 2x - 1$ 9. $4y^4 + 5y^2 - 2y$ 10. $-3a^2 b + ab^3$ 11. $b^3 - 4b$ 12. $-2a^2 - 4a$

Objective 5.4B

1. $x - 7$ 2. $x - 8$ 3. $5x$ 4. $x + 12$ 5. $x^2 + 5x + 7 + \dfrac{9}{x - 2}$ 6. $2x - 1 - \dfrac{2}{3x + 2}$

7. $6x + 2 - \dfrac{7}{2x + 1}$ 8. $5x - 2 + \dfrac{5}{3x + 1}$ 9. $2x^2 - 1 - \dfrac{10}{3x^2 - 2}$ 10. $9x^2 - 6x + 4$

11. $2x^2 - 2x - 11 - \dfrac{15}{2x - 1}$ 12. $2x^2 - 3x - 2$ 13. $x^2 - 2x - 3$ 14. $2x^2 - 3x + 2$ 15. $x^2 - 2x - \dfrac{12}{x - 4}$

16. $x^2 - 3x + 6$ 17. $3x^3 - 2x^2 - x + 2$ 18. $2x^3 - x^2 - x + 1$

Objective 5.4C

1. $2x - 5$ 2. $4x - 2 + \dfrac{1}{x + 3}$ 3. $2x - 7$ 4. $3x + 5 + \dfrac{5}{x - 4}$ 5. $4x + 4 + \dfrac{1}{x - 1}$

6. $4x - 12 + \dfrac{28}{x + 3}$ 7. $2x^2 - x + 7$ 8. $3x^2 + 5x - 3$ 9. $3x^2 + 9x + 28 + \dfrac{69}{x - 3}$ 10. $x^2 - 6x + 12 - \dfrac{14}{x + 2}$

11. $4x^2 - x - 2$ 12. $3x^2 + 11x + 5 + \dfrac{7}{x - 1}$ 13. $2x^2 + 5x + 19 + \dfrac{105}{x - 5}$ 14. $4x^2 - 3x + 1$ 15. $3x^3 - 12x^2 + 28x - 59 + \dfrac{114}{x + 2}$

16. $3x^3 - x^4 - 4x + 11 + \dfrac{31}{x - 3}$ 17. $3x^3 + 6x^2 + 11x + 22 + \dfrac{47}{x - 2}$ 18. $x^3 + 3x^2 + 5x + 15 + \dfrac{27}{x - 3}$

Objective 5.4D

1. 16   2. -25 3. -143 4. 193   5. 6   6. 19   7. 280

8. -4 9. 600   10. -19 11. 5   12. 4   13. 0   14. 579

15. 7   16. -2

155

Objective 5.5A

1. $3b(5b+2)$ **2.** $x^2(3x-4)$ **3.** $3(a^2-6b^3)$ **4.** $2(8x^2+9y^2)$ **5.** $x(x^3+x-2)$ **6.** $x(3x^2-2x+5)$

7. $x^2(2x^3-x^2-4)$ **8.** $x^2(x^2-x+4)$ **9.** $3(x^2+3x+6)$ **10.** $5(2x^2-3x+4)$

11. $4(5x^2-2x+7)$ **12.** $7(4x^2-3x+2)$ **13.** $4b^2(2-3b+4b^2)$ **14.** $6b^2(6b^2-2b+1)$

15. $x^{2n}(1-x^n)$ **16.** $a^{5n}(a^n-1)$ **17.** $x^n(x^{3n}-x^n+3)$ **18.** $x^n(x^{2n}+1)$ **19.** $a^{3n}(a^{3n}-1)$ **20.** $a^4(a^n-1)$

21. $5x^2y(3y-4x+2)$ **22.** $8x^2y^2(5xy-6y+7)$ **23.** $12x^2y(1+2y+3y^2)$ **24.** $a^n(a^{n+2}-a^{n+1}+2)$

Objective 5.5B

1. $(x+y)(2+a)$ **2.** $(x-4)(a-b)$ **3.** $(a-c)(b+d)$ **4.** $(x+1)(x+5)$ **5.** $(x-4)(x+7)$

6. $(x+6)(x-2)$ **7.** $(x+8)(x-3)$ **8.** $(a+6)(b-3)$ **9.** $(a+c)(d-e)$

10. $(b+4)(a^2+1)$ **11.** $(5+y)(2-x^2)$ **12.** $(4+b)(3-a^2)$ **13.** $(6+a)(2+a^2)$

14. $(5-a)(4+a^2)$ **15.** $(2a-b)(x^2+3y)$ **16.** $(2x-y)(2a^2+3b)$ **17.** $(x^n-4)(a^n-3)$

18. $(x^n+3)(a^n+2)$

Objective 5.6A

1. $(x-2)(x-5)$ **2.** $(x+9)(x+2)$ **3.** $(a+8)(a+1)$ **4.** $(a-8)(a+7)$ **5.** $(b+7)(b-4)$ **6.** $(a+5)(a+1)$

7. $(a-10)(a+2)$ **8.** $(x+7)(x+5)$ **9.** $(x+9)(x-7)$ **10.** $(a-9)(a-4)$

11. $(a-9)(a-7)$ **12.** $(x+10)(x-1)$ **13.** $(a-9)(a-5)$ **14.** $(a-17)(a+2)$

15. $(a+13)(a-2)$ **16.** $(a+9)(a-2)$ **17.** $(x-9)(x-3)$ **18.** $(x+7)(x+8)$

19. $(x+10)(x-9)$ **20.** $(a-9)(a-3)$ **21.** $(x+10)(x+4)$ **22.** $(x+12)(x+3)$

23. $(b-12)(b+4)$ **24.** $(x+8)(x-1)$ **25.** $(a-b)(a-4b)$ **26.** $(a-8b)(a-b)$

27. $(x+10y)(x-5y)$

Objective 5.6B

1. $(2x+1)(x+2)$ **2.** $(2x+3)(x-4)$ **3.** $(3x-2)(x+5)$ **4.** $(3a+5)(a-1)$ **5.** $(2a-1)(a+6)$

6. $(3y-8)(y+2)$ **7.** $(3b-2)(2b-3)$ **8.** $(2x+3)(x-1)$ **9.** $(2x-1)(5x+3)$

10. $(6x-1)(x+5)$ **11.** $(2x+5)(x-1)$ **12.** $(5x-1)(x+2)$ **13.** $(4x+1)(x+3)$

14. $(a-1)(3a+4)$ **15.** $(3x-1)(5x+2)$ **16.** $(10x-1)(x-8)$ **17.** $(4x-3)(3x+2)$

18. $(2x-1)(2x+3)$ **19.** $(4x-3)(x+1)$ **20.** $(3y-2)(4y+5)$ **21.** $(2x-y)(3x-4y)$

22. $(3a+5b)(a+8b)$ **23.** $(6a-b)(a+5b)$ **24.** $(2x+3y)(4x-y)$ **25.** Nonfactorable over the integers

Objective 5.7A

1. $(x+3)(x-3)$ **2.** $(x+6)(x-6)$ **3.** $(3x+1)(3x-1)$ **4.** $(3x+4)(3x-4)$ **5.** $(6x+8)(6x-8)$

6. $(ab+11)(ab-11)$ **7.** Nonfactorable **8.** Nonfactorable **9.** $(4+ab)(4-ab)$

10. $(7+xy)(7-xy)$ **11.** $(6a+5b^2)(6a-5b^2)$ **12.** $(a^n+3)(a^n-3)$ **13.** $(b^n+6)(b^n-6)$ **14.** $(a+9)^2$

15. Nonfactorable **16.** Nonfactorable **17.** Nonfactorable **18.** Nonfactorable **19.** $(x+2y)^2$

20. $(4xy+25)(xy+1)$ **21.** $(5a-2b)^2$ **22.** $(2a-5b)^2$ **23.** $(x^n+4)^2$ **24.** $(y^n-10)^2$

Objective 5.7B

1. $(x-2)(x^2+2x+4)$ **2.** $(x-4)(x^2+4x+16)$ **3.** $(3x-1)(9x^2+3x+1)$ **4.** $(2x+3)(4x^2-6x+9)$

5. $(x+2y)(x^2-2xy+4y^2)$ **6.** $(x+6y)(x^2-6xy+36y^2)$ **7.** $(m-n)(m^2+mn+n^2)$

8. $(5x-y)(25x^2+5xy+y^2)$ **9.** $(6a+1)(36a^2-6a+1)$ **10.** $(1-6a)(1+6a+36a^2)$

11. $(4x-3y)(16x^2+12xy+9y^2)$ **12.** $(3x+2y)(9x^2-6xy+4y^2)$ **13.** $(xy+2)(x^2y^2-2xy+4)$

14. $(3xy+4)(9x^2y^2-12xy+16)$ **15.** Nonfactorable **16.** Nonfactorable **17.** Nonfactorable

18. $(3x-5)(9x^2-15x+25)$ **19.** $(5-x)(25+5x+x^2)$ **20.** $(4+x)(16-4x+x^2)$ **21.** Nonfactorable

22. $(a+b-2)[(a+b)^2+(a+b)+4]$ **23.** $-b[a^2+a(a+b)+(a+b)^2]$ **24.** Nonfactorable

25. $(x^n-y^n)(x^{2n}+x^ny^n+y^{2n})$ **26.** $(y^n+4)(y^{2n}-4y^n+16)$ **27.** $(a^n-5)(a^{2n}+5a^n+25)$

Objective 5.7C

1. $(xy-10)(xy+2)$ **2.** $(xy-8)(xy-5)$ **3.** $(ab+7)(ab+4)$ **4.** $(ab-7)(ab-8)$

5. $(x^2-8)(x^2-2)$ **6.** $(y^2-14)(y^2+3)$ **7.** $(b^2-7)(b^2-3)$ **8.** $(b^2+9)(b^2+3)$

9. $(a^2-12)(a^2+5)$ **10.** $(a^2b^2-2)(a^2b^2+12)$ **11.** $(ab-3)(ab+11)$ **12.** $(a^n-5)(a^n+3)$

13. $(2xy-5)(xy+4)$ **14.** $(3xy-4)(xy+2)$ **15.** $(3ab-2)(ab-5)$ **16.** $(2a^n+3)(d^n+3)$

17. $(2a^n+1)(a^n-5)$ **18.** $(3a^n+2)(2a^n+1)$

Objective 5.7D

1. $4(x-1)^2$ **2.** $5(x+2)(x-2)$ **3.** $6(x+1)(x-1)$ **4.** $4(2x+3)(2x-3)$ **5.** $(x^2+9)(x+3)(x-3)$

6. Nonfactorable **7.** $3x^2(2x+5)(2x-5)$ **8.** $3a(1-a)(1+a+a^2)$ **9.** $a^3(ab+3)(a^2b^2-3ab+9)$

10. $3x^2(x+8)(x-5)$ **11.** $x^2(x-7)(x+4)$ **12.** $(a^2+9)(a+2)(a-2)$ **13.** $(a^2+1)(a+3)(a-3)$ **14.** $3a^2(a-4)^2$

15. $(a+1)(a-1)(a+2)(a-2)$ **16.** $(x^2+4)(x+2)(x-2)$ **17.** $5(2x-3)(x+1)$ **18.** $6(y+3)(y+2)$

19. $3xy(x-2y)(5x+4y)$ **20.** $2x^2(2x+3y)(4x-5y)$ **21.** $a^2(a^n-4)^2$ **22.** $x(x^n+2)^2$ **23.** $x^n(x-2)(2x-3)$

24. $a^n(2a-1)(a+3)$

Objective 5.8A

1. $-6, 4$ **2.** $-\frac{4}{3}, 7$ **3.** $0, \frac{2}{5}, \frac{8}{3}$ **4.** $-\frac{1}{3}, 0, 6$ **5.** $0, 3$ **6.** $0, 3$ **7.** $-\frac{2}{5}, \frac{2}{5}$

8. $-3, 3$ **9.** $-3, 4$ **10.** $-3, 5$ **11.** $-2, -1, 0$ **12.** $-2, 2$ **13.** $-3, 1, 3$ **14.** $-2, \frac{7}{3}, 2$

15. $-1, \frac{4}{3}, 1$ **16.** $-4, \frac{5}{3}$

Objective 5.8B

1. $-7, 6$ **2.** $-10, 9$ **3.** length: 33 cm; width: 6 cm **4.** base: 8 cm; height: 22 cm **5.** 25 cm
6. length: 41 ft; width: 7 ft **7.** 3 s **8.** 1.5 s

CHAPTER 6

Objective 6.1A

1. $-\dfrac{7}{2}$ **2.** $\dfrac{2}{17}$ **3.** $-\dfrac{1}{2}$ **4.** $-\dfrac{2}{9}$ **5.** $\dfrac{3}{29}$ **6.** $\dfrac{6}{5}$ **7.** $-\dfrac{5}{19}$

8. $-\dfrac{1}{3}$ **9.** $\{x \mid x \neq -3\}$ **10.** $\{x \mid x \neq -3,\, 4\}$ **11.** $\{x \mid x \neq -1\}$ **12.** $\{x \mid x \neq 7\}$ **13.** $\{x \mid x \neq -3,\, 3\}$

14. $\{x \mid x \neq -2,\, 5\}$ **15.** $\{x \mid x \neq -4\}$ **16.** $\{x \mid x \neq -2,\, 3\}$

Objective 6.1B

1. $1 + 3x$ **2.** $3x - 1$ **3.** $5x$ **4.** $\dfrac{4x}{3}$ **5.** $-\dfrac{3}{x}$ **6.** $2x^2 - x + 3$ **7.** $\dfrac{x^n - 3}{4}$

8. $\dfrac{a-3}{a+7}$ **9.** $\dfrac{x-5}{x+5}$ **10.** $\dfrac{x^n + y^n}{x^n - y^n}$ **11.** $\dfrac{3}{a^n - 3}$ **12.** $-\dfrac{x+4}{2x-1}$ **13.** $\dfrac{3x+2}{3x-2}$ **14.** $\dfrac{x-4y}{x+4y}$

15. $\dfrac{x+9}{x-3}$ **16.** $\dfrac{9x^2 - 3xy + y^2}{3x + y}$ **17.** $\dfrac{2xy + 1}{2xy + 5}$ **18.** $\dfrac{x-3}{a-b}$ **19.** $\dfrac{x^2 - 8}{x^2 + 8}$ **20.** $\dfrac{x^2 + 4}{x^2 + 3}$

21. $\dfrac{xy - 4}{xy - 7}$ **22.** $\dfrac{a^n + 4}{a^n + 6}$ **23.** $\dfrac{a^n + 2}{a^n + 5}$

Objective 6.2A

1. $\dfrac{ab^2}{x}$ **2.** $\dfrac{ax}{b}$ **3.** $\dfrac{2x-1}{y^2(x+1)}$ **4.** $\dfrac{x^2(x+2)}{y^2(x+1)}$ **5.** $\dfrac{x-3}{x-1}$ **6.** $\dfrac{x+4}{x+2}$ **7.** $\dfrac{x+6}{x-6}$

8. $-\dfrac{2x-5}{2x+5}$ **9.** $\dfrac{4(x+3)}{3x(x-1)}$ **10.** 1 **11.** $\dfrac{3x-5}{2x-5}$ **12.** $\dfrac{x+5}{x-5}$ **13.** $\dfrac{3x}{x-3}$ **14.** $\dfrac{x-5}{x+5}$

15. $\dfrac{x^n + 2}{x^n - 2}$ **16.** $(x^n - 2)^2$ **17.** $x + 2y$ **18.** $\dfrac{(x-1)(3x-2)}{(3x+1)(x+5)}$

Objective 6.2B

1. $\dfrac{y}{8ax^2}$ **2.** $\dfrac{b^3}{axy^2}$ **3.** $\dfrac{14}{15}$ **4.** $\dfrac{2(x-y)(x-y^2)}{3x^2 y^2}$ **5.** $\dfrac{x^2 y(2x+1)}{x-1}$ **6.** $\dfrac{xy(x-1)}{x+3}$

7. $\dfrac{x-2}{x+6}$ **8.** $-\dfrac{x+6}{x-4}$ **9.** 2 **10.** $-\dfrac{2x-9}{x}$ **11.** $\dfrac{x-2}{x-4}$ **12.** $\dfrac{2(x+3)}{x-2}$ **13.** 2

14. $-\dfrac{x+2}{x+3}$ **15.** $\dfrac{x^n - 5}{x^n - 4}$ **16.** $\dfrac{(x^n + 3)^2}{(x^n - 3)(x^n - 2)}$ **17.** $x + 5$ **18.** $(x-4)(x+1)$

Objective 6.3A

1. $\dfrac{3x^2-9x}{6x^2(x-3)}$, $\dfrac{2x-6}{6x^2(x-3)}$

2. $\dfrac{2x^3-3x^2}{4x^3(2x-1)}$, $\dfrac{6x-3}{4x^3(2x-1)}$

3. $\dfrac{3x-2}{4x^2-12x}$, $\dfrac{-16x^3+48x^2}{4x^2-12x}$

4. $\dfrac{5x-4}{2x(x-4)}$, $\dfrac{6x^3-24x^2}{2x(x-4)}$

5. $\dfrac{30x-18}{(5x+3)(5x-3)}$, $\dfrac{-40x-24}{(5x+3)(5x-3)}$

6. $\dfrac{5x}{(x+2)(x-2)}$, $\dfrac{x^2+5x+6}{(x+2)(x-2)}$

7. $\dfrac{6x}{3(3+x)(3-x)}$, $\dfrac{12x+x^2}{3(3+x)(3-x)}$

8. $\dfrac{15}{6(x+2y)(x-2y)}$, $\dfrac{6x+12y}{6(x+2y)(x-2y)}$

9. $\dfrac{15x}{5(x+5)(x-5)}$, $\dfrac{x^2+3x-10}{5(x+5)(x-5)}$

10. $\dfrac{2x^2+2x}{(x+1)^2(x-1)}$, $\dfrac{3x^3-3x}{(x+1)^2(x-1)}$

11. $\dfrac{x^2-1}{(x-2)(x^2+2x+4)}$, $\dfrac{3x-6}{(x-2)(x^2+2x+4)}$

12. $\dfrac{x-4}{(4-x)(16+4x+x^2)}$, $\dfrac{12-3x}{(4-x)(16+4x+x^2)}$

13. $\dfrac{7x^2-14x}{(x+3)(x-2)^2}$, $\dfrac{-4x^2-12x}{(x+3)(x-2)^2}$

14. $\dfrac{2x^2+6x}{(x-1)(x+3)^2}$, $\dfrac{3x^2-3x}{(x-1)(x+3)^2}$

15. $\dfrac{-3x^2+9x}{(2x+1)(x-2)(x-3)}$, $\dfrac{2x^2-4x}{(2x+1)(x-2)(x-3)}$

16. $\dfrac{10x^2+25}{(2x-5)(2x+5)(2x+3)}$, $\dfrac{-6x^2+15x}{(2x-5)(2x+5)(2x+3)}$

17. $\dfrac{4}{(2x-1)(x+3)}$, $\dfrac{-3x^2-9x}{(2x-1)(x+3)}$, $\dfrac{4x^2-1}{(2x-1)(x+3)}$

18. $\dfrac{6}{(3x-2)(x-4)}$, $\dfrac{-3x^2-2x}{(3x-2)(x-4)}$, $\dfrac{x^2-5x+4}{(3x-2)(x-4)}$

19. $\dfrac{2x^2+4x}{(x-5)(x+2)}$, $\dfrac{4x-20}{(x-5)(x+2)}$, $\dfrac{-x-3}{(x-5)(x+2)}$

20. $\dfrac{3x^2+9x}{(x-2)(x+3)}$, $\dfrac{-x+2}{(x-2)(x+3)}$, $\dfrac{-x+2}{(x-2)(x+3)}$

21. $\dfrac{3x^n+6}{(x^n+2)^2(x^n-2)}$, $\dfrac{6x^n-12}{(x^n+2)^2(x^n-2)}$

22. $\dfrac{x-3}{(x^n+2)(x^n-1)}$, $\dfrac{5x^{n+1}+10x}{(x^n+2)(x^n-1)}$

Objective 6.3B

1. 0

2. $\dfrac{1}{5x^2}$

3. $\dfrac{1}{x+1}$

4. $\dfrac{1}{x+5}$

5. $\dfrac{1}{x+y}$

6. $\dfrac{1}{x+2}$

7. $\dfrac{5-4xy-35y}{15xy^2}$

8. $\dfrac{18b-15a-2}{24a^2b^2}$

9. $\dfrac{11}{30x}$

10. $\dfrac{a^2+10a-4}{a(a-2)}$

11. $\dfrac{2(x^2+8x+2)}{(x+5)(x+4)}$

12. $\dfrac{10-2x}{(x+5)^2}$

13. $\dfrac{x+1}{x+5}$

14. $\dfrac{-2x+4}{x^2+8x+16}$

15. $\dfrac{x^2-9x}{(x-4)(x-5)}$

16. $\dfrac{-5a^2+38a}{(a+4)(a-1)}$

17. $\dfrac{4}{(x+4)(x+3)}$

18. $-\dfrac{x^{2n}+x^n+9}{(x^n-3)(x^n+4)}$

19. $\dfrac{5x+14}{9-4x^2}$

20. $\dfrac{x^2-40x+83}{5x^2-20}$

Objective 6.4A

1. $\dfrac{x}{3x-1}$

2. $\dfrac{4-x}{x}$

3. $-\dfrac{a}{a+5}$

4. $\dfrac{6-a}{a}$

5. 1

6. $\dfrac{x}{4(4-x)}$

7. $\dfrac{x+4}{x+6}$

8. $\dfrac{3-2x}{2-3x}$

9. $\dfrac{3x-1}{x-3}$

10. $\dfrac{2(x+8)}{5x-3}$

11. $\dfrac{2x-8}{2x^2-3x-24}$

12. $\dfrac{3}{3x-4}$

13. $\dfrac{x-4}{x+6}$

14. $\dfrac{x-1}{x+2}$

15. $-\dfrac{2(a+3)}{4a-3}$

16. $-\dfrac{3(a-4)}{8(a+2)}$

17. $-\dfrac{4x}{x^2+4}$

18. $-\dfrac{4}{x}$

Objective 6.5A

1. 12	**2.** 36	**3.** 10	**4.** 25	**5.** 6	**6.** 4	**7.** −19
8. −18	**9.** 7	**10.** −3	**11.** $-\dfrac{5}{3}$	**12.** 6	**13.** −4	**14.** 1
15. 5	**16.** 4	**17.** 2	**18.** $-\dfrac{1}{2}$			

Objective 6.5B

1. $90,000 **2.** $115.20 **3.** 228,000 people **4.** 21 DVDs **5.** 76 ft by 96 ft **6.** $93.75
7. $\dfrac{5}{8}$ oz **8.** 15 oz **9.** 225 shares **10.** $7000

Objective 6.6A

1. −3	**2.** $\dfrac{5}{2}$	**3.** 1	**4.** −3	**5.** −2	**6.** −4	**7.** 3
8. −2	**9.** −5	**10.** 12	**11.** 5	**12.** no solution	**13.** −3	**14.** 2
15. −6	**16.** −9	**17.** 4	**18.** −8			

Objective 6.6B

1. 36 min	**2.** 18 min	**3.** 12 min	**4.** 12 h	**5.** 24 h	**6.** 30 h	**7.** 6 min
8. 5 h	**9.** 24 min	**10.** 4.8 h				

Objective 6.6C

1. 48 mph **2.** corporate jet: 435 mph; commercial jet: 540 mph **3.** 50 mph **4.** 400 mph **5.** 50 mph
6. 55 mph **7.** 28 mph **8.** 60 mph **9.** freight train: 40 mph; passenger train: 55 mph **10.** 60 mph

Objective 6.7A

1. 21 lb **2.** 10 lb/in^2 **3.** 234.4 ft **4.** 128 ft **5.** 15 rpm **6.** 105 lb/in^2 **7.** 192 lumens
8. 16 ft **9.** 135 times/min **10.** 17.5 amps

CHAPTER 7

Objective 7.1A

1. 5	**2.** 3	**3.** 64	**4.** $\dfrac{1}{4}$	**5.** Not a real number	**6.** −3	**7.** $\dfrac{7}{4}$
8. $\dfrac{1}{x^{2/5}}$	**9.** $a^{5/3}$	**10.** $x^{3/4}$	**11.** a^{4n}	**12.** a	**13.** $\dfrac{1}{x^{1/5}}$	**14.** $a^3 b^6$
15. $\dfrac{x^2}{y^3}$	**16.** $x^2 y^{12}$	**17.** $x^{4/3} - x^{1/3}$	**18.** a	**19.** x^{2n}	**20.** $a^{n/4}$	**21.** $\dfrac{1}{x^{5/3}}$
22. $\dfrac{1}{a^{2n}}$	**23.** x	**24.** $\dfrac{1}{x}$	**25.** $\dfrac{1}{a^4}$	**26.** $\dfrac{1}{x^6}$	**27.** $\dfrac{1}{x^9}$	

Objective 7.1B

1. $\sqrt[3]{2}$ 　　2. $\sqrt{7}$ 　　3. $\sqrt[5]{b^3}$ 　　4. $\sqrt[4]{128x^7}$ 　　5. $-9\sqrt[3]{a^2}$ 　　6. $\sqrt[3]{ab^2}$ 　　7. $\sqrt[5]{(x^3y^2)^3}$

8. $\sqrt[3]{(a^2b^2)^2}$ 　　9. $\sqrt[3]{3a-2b}$ 　　10. $\sqrt[3]{3x-4}$ 　　11. $\sqrt[3]{(2x-1)^2}$ 　　12. $\dfrac{1}{\sqrt[4]{x^3}}$ 　　13. $15^{1/2}$ 　　14. $y^{1/3}$

15. $\left(5y^5\right)^{1/4}$ 　　16. $-\left(2x^3\right)^{1/2}$ 　　17. $-\left(3x^5\right)^{1/3}$ 　　18. $-\left(2x^7\right)^{1/4}$ 　　19. $-\left(4x^4\right)^{1/3}$ 　　20. $2xy^{2/3}$ 　　21. $3xy^{1/4}$

22. $\left(a^2-8\right)^{1/2}$ 　　23. $\left(2-x^2\right)^{1/2}$ 　　24. $\left(x^2+y^2\right)^{1/2}$

Objective 7.1C

1. x^6 　　2. x^5 　　3. $-a^2$ 　　4. x^2y^3 　　5. $-x^4y^2$ 　　6. $-a^2b^5$ 　　7. $-ab^3$

8. x^2y^5 　　9. x^3y^5 　　10. $-a^3b^4$ 　　11. $12x^6y^8$ 　　12. $3a^2b^5$ 　　13. $9x^3y^2$ 　　14. $4x^2$

15. $5x^4y$ 　　16. $-x^4y^7$ 　　17. $-2a^6b^3$ 　　18. $-a^2b^3$ 　　19. x^4y^6 　　20. x^3y^5 　　21. $2a^4b^5$

22. $5a^2b^3$ 　　23. $-ab^3$ 　　24. $3x^2y^3$

Objective 7.2A

1. $x^2y^3z^3\sqrt{z}$ 　2. $7a^2b^3\sqrt{2}$ 　3. $5y^3z^4\sqrt{2xz}$ 　4. $5xy^2z^3\sqrt{3x}$ 　5. Not a real number 　　6. $-xy^3\sqrt{x}$

7. $x^2y^3z^4\sqrt{xyz}$ 　8. $x^3y^4z^2$ 　　9. x^3z^6 　　10. $8y^3z^5\sqrt{3xy}$ 　11. $8xy^2z^3\sqrt{2xy}$ 　　12. $2x^2y^3z^3\sqrt[3]{z}$

13. $-4xy^3$ 　　14. Not a real number 　　15. Not a real number 　　16. $ab^2\sqrt[3]{a^2b}$ 　17. $a^2b^2\sqrt[3]{b^2}$ 　18. $-2x^2y\sqrt[3]{y}$

19. $-4x^2y^3$ 　　20. $ab^2c^2\sqrt[3]{a^2c}$ 　　21. $a^4b^3c^2\sqrt[3]{c^2}$ 　　22. $a^3b^2c^2\sqrt[3]{b}$ 　23. $2xy\sqrt[4]{x^3y}$ 　24. $2x^3y^2\sqrt[4]{y}$

Objective 7.2B

1. $-4\sqrt{x}$ 　　2. $14\sqrt{x}$ 　　3. $-\sqrt{5}$ 　　4. $\sqrt{3}$ 　　5. $\sqrt{2x}$ 　　6. $11\sqrt{3x}$ 　　7. $\sqrt{6a}$

8. $9\sqrt{3a}$ 　　9. $13x\sqrt{3x}$ 　　10. $2x\sqrt{3xy}$ 　　11. $-7a\sqrt[3]{2a^2}$ 　12. $14a^2b^2\sqrt{2ab}$ 　　13. $3ab^4\sqrt{ab}$

14. -1 　　15. $6\sqrt[3]{3}$ 　　16. $-9a\sqrt[3]{2a}$ 　　17. $20y\sqrt{3y}$ 　　18. $6a\sqrt{a}$ 　　19. $ab\sqrt[4]{ab}$ 　　20. $6xy\sqrt[3]{2x}$

21. $-13a^2\sqrt{2a}$ 　22. $-7x^2y\sqrt[4]{3xy}$

Objective 7.3A

1. 4 　　2. $4\sqrt{3}$ 　　3. 14 　　4. $4\sqrt[3]{4}$ 　　5. x^2y^4 　　6. $3x^2y^3\sqrt[3]{2}$ 　　7. $4ab^3\sqrt[3]{2a}$

8. 8 　　9. -4 　　10. $x+\sqrt{3x}$ 　　11. $x-\sqrt{2x}$ 　　12. $8x\sqrt{2}-6\sqrt{2x}$ 　　13. $2x+6\sqrt{2x}+9$

14. $x+4\sqrt{x}+4$ 　　15. $3x-12\sqrt{3x}+36$ 　　16. $48x^3y^3\sqrt{3}$ 　17. $900x^3y^4$ 　18. $6a^2b^2$ 　19. $3a^2b^3\sqrt[3]{6a^2}$

20. $\sqrt{2}$ 　　21. $-2\sqrt{3}$ 　　22. $x-9$ 　　23. $x-16$

Objective 7.3B

1. $4x$ **2.** $a^2b\sqrt{2}$ **3.** $3ab^2$ **4.** $\dfrac{\sqrt{3}}{3}$ **5.** $\dfrac{\sqrt{5}}{5}$ **6.** $\dfrac{\sqrt{6x}}{6x}$ **7.** $\dfrac{3\sqrt{2y}}{2y}$

8. $\dfrac{3\sqrt{x}}{x}$ **9.** $\dfrac{2\sqrt{6a}}{3a}$ **10.** $\dfrac{\sqrt{5x}}{x}$ **11.** $\dfrac{\sqrt{2a}}{a}$ **12.** $\dfrac{2\sqrt[3]{9}}{3}$ **13.** $\sqrt[3]{3}$ **14.** $\dfrac{x\sqrt{2xy}}{2y}$

15. $\dfrac{b\sqrt{6}}{4a^2}$ **16.** $\dfrac{x^3\sqrt{15}}{5}$ **17.** $\dfrac{3\sqrt{5}-3}{4}$ **18.** $\dfrac{9}{4}+\dfrac{3}{4}\sqrt{5}$ **19.** $\dfrac{2\sqrt{x}+8}{x-16}$ **20.** $\dfrac{5\sqrt{x}+25}{25-x}$ **21.** $\dfrac{7-2\sqrt{10}}{3}$

22. 11 **23.** 37 **24.** $x+1$ **25.** $x+1$ **26.** $\dfrac{9}{2}x\sqrt{2}-12x$ **27.** $2x-4$

Objective 7.4A

1. 9   **2.** 16   **3.** 49   **4.** 36   **5.** 8   **6.** 64   **7.** 216
8. 12   **9.** 32   **10.** 4   **11.** −72   **12.** No solution   **13.** 16   **14.** 3
15. 7   **16.** 6   **17.** 4   **18.** 2   **19.** 20   **20.** 8   **21.** −7
22. 7   **23.** 2   **24.** −5   **25.** 0

Objective 7.4B

1. 12 ft   **2.** 18.7 ft   **3.** 12.25 ft   **4.** 14.88 ft   **5.** 87.89 ft   **6.** 1600 ft   **7.** 152 m
8. 200 ft   **9.** 3.25 ft   **10.** 8.31 ft

Objective 7.5A

1. $7i$ **2.** $2i\sqrt{2}$ **3.** $2i\sqrt{3}$ **4.** $10i\sqrt{2}$ **5.** $8i\sqrt{2}$ **6.** $12i$ **7.** $6i\sqrt{3}$
8. $6i\sqrt{5}$ **9.** $3+3i$ **10.** $4+5i$ **11.** $6+4i$ **12.** $7+10i$ **13.** $4\sqrt{3}-3i\sqrt{3}$ **14.** $5\sqrt{2}+3i\sqrt{2}$
15. $2\sqrt{6}-6i$ **16.** $5\sqrt{2}-7i\sqrt{2}$ **17.** $5\sqrt{6}-3i\sqrt{3}$ **18.** $6\sqrt{2}-8i\sqrt{2}$

Objective 7.5B

1. $11-6i$ **2.** $11-9i$ **3.** $12+3i$ **4.** $12-i$ **5.** $13-7i$ **6.** $-6-2i$ **7.** $-6+i$
8. $1-75i$ **9.** $82-64i$ **10.** $-72-65i$ **11.** $9-i$ **12.** $8-3i$ **13.** $6-i$
14. $7-i$ **15.** 0 **16.** $-3-3i$ **17.** $2-i$ **18.** $5+2i$

Objective 7.5C

1. 40 **2.** -15 **3.** $-160i$ **4.** -32 **5.** -9 **6.** -8 **7.** $-4\sqrt{3}$
8. $-3+15i$ **9.** $-12+8i$ **10.** $17-i$ **11.** $9+13i$ **12.** $24-10i$ **13.** $11+3i$ **14.** $2-11i$
15. 40 **16.** 34 **17.** $13-i$ **18.** $3-21i$ **19.** $6+13i$ **20.** $36+3i$ **21.** $\dfrac{5}{3}+\dfrac{5}{3}i$

22. $\dfrac{11}{16}-\dfrac{7}{16}i$ **23.** $\dfrac{3}{4}+\dfrac{1}{4}i$ **24.** $\dfrac{11}{5}+\dfrac{3}{5}i$

Objective 7.5D

1. $-4i$ **2.** $-\dfrac{2}{3}i$ **3.** $-\dfrac{3}{4}i$ **4.** $2i+3$ **5.** $3i-\dfrac{2}{5}$ **6.** $\dfrac{20}{17}+\dfrac{5}{17}i$ **7.** $\dfrac{12}{5}-\dfrac{4}{5}i$

8. $\dfrac{25}{29}-\dfrac{10}{29}i$ **9.** $\dfrac{16}{17}+\dfrac{4}{17}i$ **10.** $\dfrac{15}{13}+\dfrac{3}{13}i$ **11.** $-i$ **12.** $-i$ **13.** $\dfrac{18}{17}+\dfrac{21}{17}i$ **14.** $\dfrac{20}{13}+\dfrac{48}{13}i$

15. $\dfrac{21}{10}-\dfrac{3}{10}i$ **16.** $4i$ **17.** $\dfrac{4+8i}{5}$ **18.** $\dfrac{4i-1}{17}$ **19.** $\dfrac{2i-1}{10}$ **20.** $\dfrac{15}{26}-\dfrac{29}{26}i$ **21.** $\dfrac{4}{5}-\dfrac{7}{5}i$

22. $1+2i$ **23.** $-\dfrac{3}{2}+\dfrac{7}{2}i$ **24.** $\dfrac{7}{5}+\dfrac{16}{5}i$

CHAPTER 8

Objective 8.1A

1. -5 and 0 **2.** -4 and 4 **3.** -8 and 8 **4.** -4 and 3 **5.** 2 **6.** -4 **7.** 0 and 3

8. -5 and 0 **9.** -3 and 5 **10.** -4 and 1 **11.** 2 and 6 **12.** -1 and 12 **13.** -1 and $\dfrac{2}{3}$ **14.** $-\dfrac{1}{2}$ and 3

15. $-\dfrac{2}{3}$ and 3 **16.** and 2 **17.** -2 and 6 **18.** -2 and 1 **19.** -5 and 3 **20.** -6 and 2 **21.** $-7b$ and $3b$

22. b and $4b$ **23.** $-c$ and $8c$ **24.** $-3a$ and $4a$ **25.** $\dfrac{c}{2}$ and $2c$ **26.** $-3a$ and $\dfrac{2a}{3}$

Objective 8.1B

1. -5 and 5 **2.** -6 and 6 **3.** $-3i$ and $3i$ **4.** $-7i$ and $7i$ **5.** -1 and 1 **6.** -3 and 3 **7.** $-5i$ and $5i$

8. $-3\sqrt{3}$ and $3\sqrt{3}$ **9.** $-5i\sqrt{2}$ and $5i\sqrt{2}$ **10.** -7 or 1 **11.** and 6 **12.** -5 or 1 **13.** -7 or 11

14. -4 and 6 **15.** $\dfrac{2}{3}$ and 0 **16.** -1 and 0 **17.** $5-i\sqrt{15}$ and $5+i\sqrt{15}$ **18.** $-4-2i\sqrt{7}$ and $-4+2i\sqrt{7}$

19. $1-3\sqrt{2}$ and $1+3\sqrt{2}$ **20.** $-2-4i$ and $-2+4i$ **21.** $3-5\sqrt{2}$ and $3+5\sqrt{2}$ **22.** $\dfrac{3}{5}+5\sqrt{3}$ and $\dfrac{3}{5}-5\sqrt{3}$

23. $\dfrac{1}{3}-3\sqrt{3}$ and $\dfrac{1}{3}+3\sqrt{3}$ **24.** $-\dfrac{1}{2}-3\sqrt{5}$ and $-\dfrac{1}{2}+3\sqrt{5}$

Objective 8.2A

1. -7 and 1 **2.** -3 and 5 **3.** 2 **4.** -4 **5.** 1 and 5 **6.** $-4-\sqrt{6}$ and $-4+\sqrt{6}$

7. -1 and 3 **8.** -5 and 2 **9.** -4 and 9 **10.** 3 and 4 **11.** 3 and 6 **12.** $-2-2\sqrt{5}$ and $-2+2\sqrt{5}$

13. $3-\sqrt{7}$ and $3+\sqrt{7}$ **14.** -1 and $\dfrac{1}{3}$ **15.** $\dfrac{1}{2}$ and 1 **16.** $-\dfrac{3}{2}$ and 4 **17.** -2 and 12 **18.** -2 and 8

19. -1 and 5 **20.** $\dfrac{2}{5}$ and $\dfrac{1}{2}$ **21.** $\dfrac{1}{2}$ and $\dfrac{3}{2}$ **22.** $-\dfrac{2}{3}$ and 6 **23.** $\dfrac{1}{3}$ and $-\dfrac{3}{2}$ **24.** $-\dfrac{1}{2}$ and $\dfrac{7}{2}$ **25.** $-\dfrac{5}{2}$ and 2

26. $\dfrac{-1-\sqrt{5}}{2}$ and $\dfrac{-1+\sqrt{5}}{2}$ **27.** 2 and 5

163

Objective 8.2B

1. −5 and 10 **2.** −5 and 8 **3.** −9 and 1 **4.** $\frac{5}{4} - \frac{\sqrt{23}}{4}i$ and $\frac{5}{4} + \frac{\sqrt{23}}{4}i$ **5.** $\frac{1}{4}$ and $-\frac{3}{2}$ **6.** $\frac{1}{3}$ and $-\frac{3}{2}$

7. $\frac{1}{2}$ and 1 **8.** $-\frac{1}{2}$ and $-\frac{9}{2}$ **9.** $-\frac{1}{2} - \frac{3}{2}i$ and $-\frac{1}{2} + \frac{3}{2}i$ **10.** $2 - i\sqrt{2}$ and $2 + i\sqrt{2}$

11. $1 - \frac{2}{3}\sqrt{3}$ and $1 + \frac{2}{3}\sqrt{3}$ **12.** $-\frac{1}{3}$ and $\frac{3}{4}$ **13.** $\frac{1}{2}$ and 5 **14.** $\frac{11}{12} - \frac{\sqrt{95}}{12}i$ and $\frac{11}{12} + \frac{\sqrt{95}}{12}i$

15. $\frac{1}{2} - \frac{\sqrt{39}}{6}i$ and $\frac{1}{2} + \frac{\sqrt{39}}{6}i$ **16.** −3 and $\frac{4}{5}$ **17.** 0 and $\frac{5}{2}$ **18.** $-\frac{4}{3}$ and 3 **19.** 0.354 and 5.646

20. 0.394 and 7.606 **21.** 0.628 and 6.372 **22.** one real **23.** two real **24.** two complex

Objective 8.3A

1. −2, −1, 1, 2 **2.** −4, −2, 2, 4 **3.** −4, −1, 1, 4 **4.** −5, −2, 2, 5 **5.** −3, −i, i, 3 **6.** −4, −2, 2, 4 **7.** 25
8. 16 **9.** 1, 16 **10.** −2, −2i$\sqrt{2}$, 2i$\sqrt{2}$, 2 **11.** −3, −2, 2, 3 **12.** $-\sqrt{3}$, −2i, 2i, $\sqrt{3}$
13. −2, $-\sqrt{2}$, $\sqrt{2}$, 2 **14.** −1, −i, i, 1 **15.** −2, −2i, 2i, 2 **16.** $-2\sqrt{2}$, $-2\sqrt{2}i$, $2\sqrt{2}i$, $2\sqrt{2}$

17. −3, $-\sqrt{7}$, $\sqrt{7}$, 3 **18.** 4 and 9 **19.** 4 and 49 **20.** 1 and 64 **21.** 1 and 64 **22.** $-2, -\frac{1}{2}, \frac{1}{2}, 2$

23. $-\frac{1}{2}$ and $\frac{1}{2}$ **24.** −1 and 1

Objective 8.3B

1. 4 **2.** 5 **3.** 5 **4.** 16 **5.** 1 and 4 **6.** 9 **7.** 3
8. 3 **9.** 4 **10.** 16 **11.** 3 and 4 **12.** 5 **13.** 8 **14.** 3
15. 7 **16.** −3 **17.** −5 and 3 **18.** 4 **19.** 9 **20.** 2 **21.** No solution
22. 2 and 6 **23.** −3 and 1

Objective 8.3C

1. −1 and 6 **2.** 3 and 4 **3.** −1 and 4 **4.** 1 and 6 **5.** −5 and 3 **6.** $-\frac{21}{4}$ and −1 **7.** $-\frac{1}{3}$

8. $-\frac{1}{5}$ and 5 **9.** 4 and 10 **10.** −1 and 7 **11.** −4 and $-\frac{9}{4}$ **12.** 0 and 4 **13.** −2 and $-\frac{2}{9}$ **14.** −4 and $\frac{3}{5}$

15. −3 and $-\frac{1}{2}$ **16.** −3 and 19 **17.** $-\frac{3}{5}$ **18.** −6 and 3 **19.** $-\frac{5}{2}$ and −1

Objective 8.4A

1. 10 ft and 4 ft **2.** 5 m, 8 m **3.** 5 and 6 **4.** −5 and −3 or 3 and 5 **5.** 2 **6.** −9 and −7 or 7 and 9 **7.** 9 h
8. smaller: 24 min; larger: 8 min **9.** older: 34 min; newer: 30 min **10.** 1st car: 48 mph; 2nd car: 60 mph

Objective 8.5A

1.   **2.**
3.
4.
5.
6.
7.
8.
9.
10.
11.
12.

CHAPTER 9

Objective 9.1A

1. 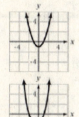 **2.** 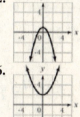 **3.** 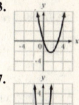 **4.**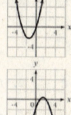

5. **6.** **7.** **8.**

Objective 9.1B

1. $(-1, 0), (1, 0)$ **2.** $(-4, 0), (4, 0)$ **3.** $(0, 0), (1, 0)$ **4.** $(0, 0), (2, 0)$ **5.** $\left(-\frac{1}{2}, 0\right), (2, 0)$ **6.** $(-6, 0), (1, 0)$

7. $\frac{1}{3}, 1$ **8.** $-3, 4$ **9.** No zeros **10.** No zeros **11.** $-1, 4$ **12.** $-3, 5$ **13.** two

14. No x-intercepts **15.** two **16.** No x-intercepts **17.** two **18.** two

19. No x-intercepts **20.** one **21.** one **22.** two **23.** two **24.** No x-intercepts

Objective 9.1C

1. minimum: -1 **2.** minimum: -4 **3.** maximum: 13 **4.** maximum: 2 **5.** minimum: -11

6. minimum: $-\frac{1}{4}$ **7.** maximum: 11 **8.** minimum: -7 **9.** maximum: $\frac{25}{4}$ **10.** minimum: -25

11. minimum: $-\frac{11}{2}$ **12.** maximum: 1 **13.** minimum: -3 **14.** minimum: $\frac{15}{4}$

15. maximum: $\frac{22}{3}$ **16.** maximum: $\frac{17}{8}$

Objective 9.1D

1. 144 ft **2.** 131 ft **3.** 5 days **4.** $187.50 **5.** 7 s **6.** 25 and 25 **7.** 15 and 15
8. -10 and 10 **9.** 14 ft by 14 ft; 196 ft^2 **10.** 8 ft by 8 ft; 64 ft^2

Objective 9.3A

1. −8 **2.** 10 **3.** −11 **4.** 11 **5.** $\frac{1}{2}$ **6.** −90 **7.** 89

8. 3 **9.** 7 **10.** −242 **11.** undefined **12.** $\frac{5}{6}$ **13.** −1 **14.** 36

15. 12 **16.** $-\frac{5}{8}$

Objective 9.3B

1. 41 **2.** 9 **3.** 51 **4.** −21 **5.** $10x - 1$ **6.** $10x + 31$ **7.** 11

8. $x^2 - 6x + 14$ **9.** $x^2 + 2$ **10.** 14 **11.** −25 **12.** −1461 **13.** 47 **14.** $8x^3 - 17$

15. $128x^3 - 480x^2 + 600x - 253$ **16.** −17

Objective 9.4A

1. No **2.** Yes **3.** Yes **4.** No **5.** No **6.** No **7.** No

8. No

Objective 9.4B

1. $f^{-1}(x) = \frac{1}{4}x - 2$ **2.** $f^{-1}(x) = x - 5$ **3.** $f^{-1}(x) = \frac{1}{3}x + 2$ **4.** $f^{-1}(x) = -x + 4$

5. $f^{-1}(x) = -4x + 8$ **6.** $f^{-1}(x) = -\frac{1}{2}x + 3$ **7.** $f^{-1}(x) = \frac{3}{2}x + 3$ **8.** $f^{-1}(x) = 4x + 24$

9. $f^{-1}(x) = \frac{1}{3}x - \frac{2}{3}$ **10.** $f^{-1}(x) = \frac{1}{5}x - \frac{1}{5}$ **11.** $f^{-1}(x) = -\frac{1}{4}x + \frac{1}{2}$ **12.** $f^{-1}(x) = \frac{1}{4}x - \frac{3}{4}$

13. Yes **14.** Yes **15.** No **16.** No

CHAPTER 10

Objective 10.1A

1. 4　　**2.** $\dfrac{1}{4}$　　**3.** 1　　**4.** 16　　**5.** $\dfrac{1}{16}$　　**6.** 64　　**7.** 3

8. 9　　**9.** $\dfrac{1}{3}$　　**10.** 1　　**11.** $\dfrac{1}{9}$　　**12.** 9　　**13.** $\dfrac{1}{3}$　　**14.** 3

15. $\dfrac{1}{81}$　　**16.** 1　　**17.** 3　　**18.** 81　　**19.** 3　　**20.** 81　　**21.** 19.683

Objective 10.1B

1. 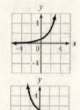　　**2.** 　　**3.** 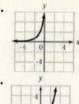　　**4.**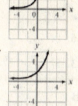

5.　　**6.**　　**7.**　　**8.**

Objective 10.2A

1. $\log_4 64 = 3$　**2.** $\log_7 49 = 2$　**3.** $\log_{64} \dfrac{1}{8} = -2$　**4.** $\log_{10} 10{,}000 = 4$　**5.** $\log_4 1 = 0$　**6.** $\log_p q = x$

7. $2^0 = 1$　**8.** $2^{-2} = \dfrac{1}{4}$　**9.** $e^c = x$　**10.** 2　**11.** -4　**12.** 2　**13.** 4

14. 4　**15.** 0　**16.** 64　**17.** 16　**18.** 216　**19.** 1　**20.** 10,000

21. 125

Objective 10.2B

1. $\log_2 x^2 y$　**2.** $\log_5 \dfrac{x^3}{y^4}$　**3.** $\log_5 x^4$　**4.** $\log_6 \dfrac{1}{x^3}$　**5.** $\log_6 \left(\dfrac{x^3}{y^4} \right)$　**6.** $\log_3 \dfrac{x^3 z^3}{y^2}$　**7.** $\log_a \dfrac{xz}{y^2}$

8. $\log_2 x^4 y^4$　**9.** $\log_5 \sqrt{xz}$　**10.** $\log_6 \sqrt[3]{\dfrac{x}{z}}$　**11.** $\log_3 \dfrac{x^5 z^3}{y^3}$　**12.** $\log_5 \sqrt{\dfrac{x^5 z}{y^3}}$　**13.** $\log_3 2 + \log_3 x$

14. $3 \log_2 x$　**15.** $\log_4 x - \log_4 y$　**16.** $3 \log_5 v - 7 \log_5 x$　**17.** $3 \log_3 x + 3 \log_3 y$　**18.** $\ln x + 3 \ln y + 2 \ln z$

19. $\dfrac{1}{2} \log_6 x + \dfrac{1}{2} \log_6 y$　**20.** $\dfrac{2}{3} \log_6 r + \dfrac{1}{3} \log_6 w$　**21.** $\dfrac{1}{2}(2 \log_5 x - \log_5 y)$　**22.** $\dfrac{1}{2} \log_5 x - \log_5 y$

Objective 10.2C

1. 1.5848　　**2.** 1.2920　　**3.** 0.7125　　**4.** 0.9358　　**5.** 1.3870　　**6.** 1.2326　　**7.** 1.7924
8. 1.9745　　**9.** 1.8981　　**10.** 2.0201　　**11.** 1.3369　　**12.** 0.7326　　**13.** 0.8617　　**14.** 0.9692
15. 1.3651　　**16.** 0.3299　　**17.** 0.5544　　**18.** 0.0937

Objective 10.3A

1.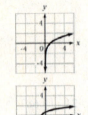

2.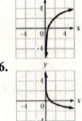

3.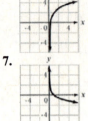

4.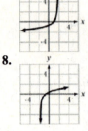

5.

6.

7.

8.

Objective 10.4A

1. −2	**2.** 1	**3.** 5	**4.** −4	**5.** 0.7782	**6.** 0.8842	**7.** −2
8. −4	**9.** 3.2656	**10.** 1.7567	**11.** 1.3802	**12.** 1.5798	**13.** −3	**14.** −2.4651
15. 4.1701	**16.** 0.7713	**17.** 2	**18.** −0.4651	**19.** $\frac{2}{3}$	**20.** 1	**21.** 3
22. −2	**23.** 3	**24.** −4				

Objective 10.4B

1. 24	**2.** 5	**3.** 14	**4.** 26	**5.** 5	**6.** 42	**7.** 5
8. −8 and 2	**9.** −4 and 9	**10.** 1	**11.** $\frac{5}{2}$	**12.** 3	**13.** No solution	**14.** $\frac{3+\sqrt{41}}{4}$
15. 2	**16.** $-\frac{2}{3}$ and 3					

Objective 10.5A

1. $12,751	**2.** 6 years	**3.** 68%	**4.** 10th week	**5.** 0.4 cm	**6.** 75%	**7.** 398 lines
8. 63,100 times						

CHAPTER 11

Objective 11.1A

1. Vertex: $\left(\dfrac{1}{2}, -\dfrac{9}{4}\right)$; Axis of symmetry: $x = \dfrac{1}{2}$

2. Vertex: $(-1, -4)$; Axis of symmetry: $x = -1$

3. Vertex: $(1, 5)$; Axis of symmetry: $x = 1$

4. Vertex: $(3, 4)$; Axis of symmetry: $x = 3$

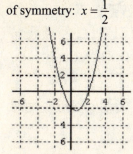
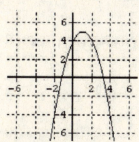
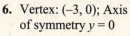
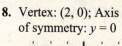

5. Vertex: $(2, -1)$; Axis of symmetry: $y = -1$

6. Vertex: $(-3, 0)$; Axis of symmetry $y = 0$

7. Vertex: $(4, 0)$; Axis of symmetry: $y = 0$

8. Vertex: $(2, 0)$; Axis of symmetry: $y = 0$

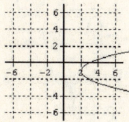
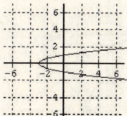
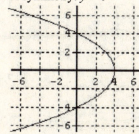
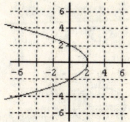

Objective 11.2A

1.

2.

3.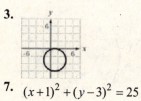

4.

5. $x^2 + (y + 4)^2 = 16$

6. $(x + 2)^2 + (y + 1)^2 = 4$

7. $(x + 1)^2 + (y - 3)^2 = 25$

8. $(x - 2)^2 + (y + 3)^2 = 16$

Objective 11.2B

1. $(x-5)^2+(y+1)^2=4$

2. $(x+4)^2+(y-2)^2=16$

3. $(x-5)^2+(y+2)^2=\dfrac{25}{4}$

4. $(x+3)^2+(y-5)^2=25$

5. $(x-3)^2+(y+4)^2=9$

6. $(x-3)^2+(y-2)^2=4$

7. $(x+3)^2+(y-2)^2=16$

8. $(x+3)^2+(y-1)^2=4$

Objective 11.3A

1.

2.

3.

4.

5.

6.

7.

8.

Objective 11.3B

1.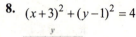

2.

3.

4.

5.

6.

7.

8.

Objective 11.4A

1. (6, 14) and (–1, 0)   **2.** (–2, 1) and (7, 19)   **3.** (–5, –13) and (3, 3)   **4.** (0, 0) and (7, 14)

5. (–1, –1) and (2, 2)   **6.** (–3, –1) and $\left(-2,-\dfrac{3}{2}\right)$ **7.** (1, 3) and (3, 1)   **8.** $\left(-\dfrac{7}{5},-\dfrac{24}{5}\right)$ and (3, 4)

9. (–2, –3) and (3, 2)   **10.** $\left(-\dfrac{8}{3},\dfrac{1}{3}\right)$ and (4, 3)   **11.** (–2, 3)   **12.** (–6, –3) and (–6, 3), (6, –3) and (6, 3)

13. (–2, 0) and (2, 0)   **14.** (–3, –2) and (–3, 2), (3, –2) and (3, 2)   **15.** (–1, 2) and $\left(-\dfrac{1}{9},\dfrac{22}{9}\right)$

16. (–1, –2) and (–1, 2), (1, –2) and (1, 2)